Custom Rifles

OF GREAT BRITAIN

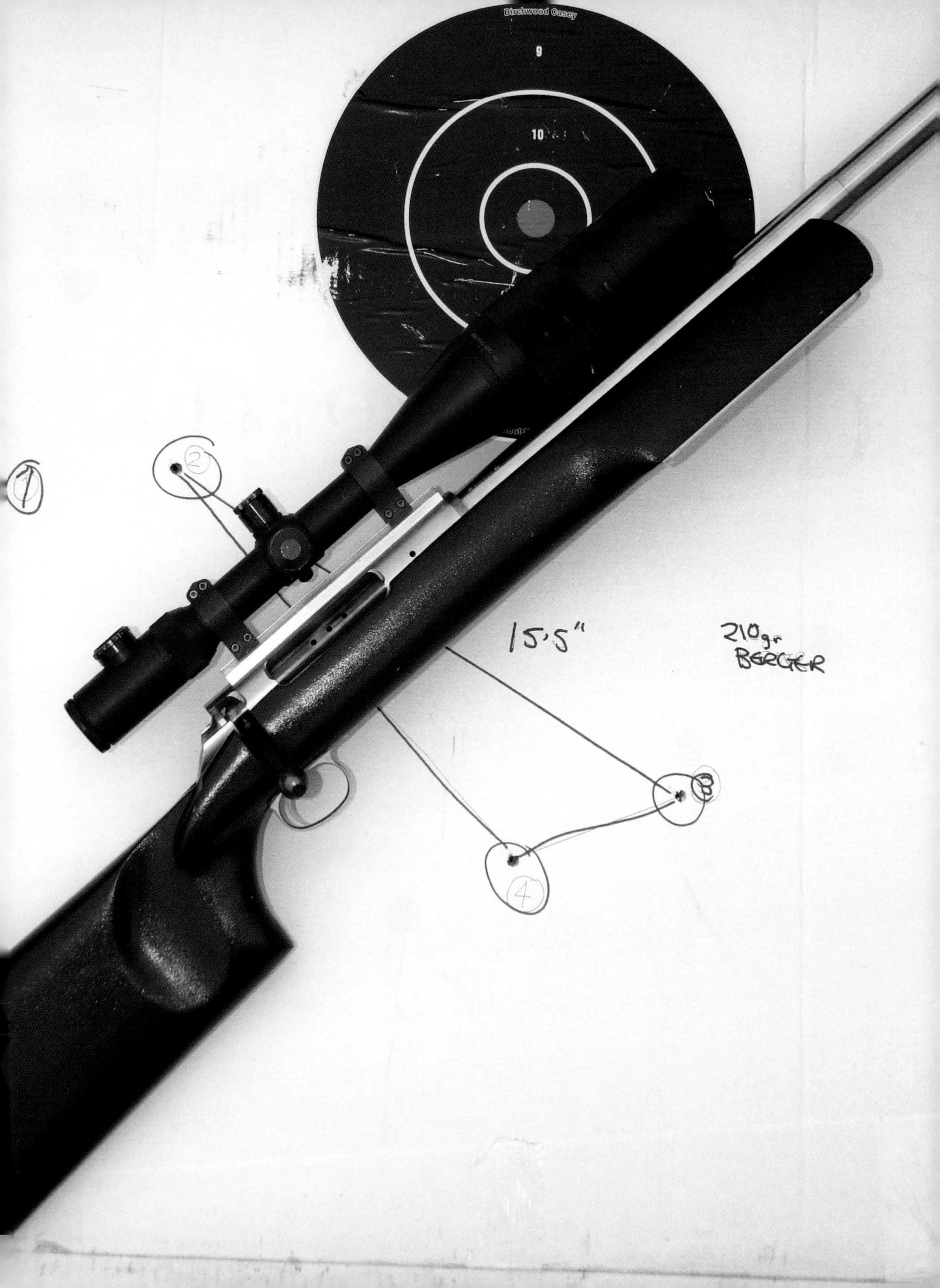
Birchwood Casey
9
10
1
2
15.5"
210gr
BERGER
3
4

Custom Rifles

OF GREAT BRITAIN

BRUCE POTTS

THE CROWOOD PRESS

First published in 2017 by
The Crowood Press Ltd
Ramsbury, Marlborough
Wiltshire SN8 2HR

www.crowood.com

British Library Cataloguing-in-Publication Data
A catalogue record for this book is available from the British Library.

ISBN 978 1 78500 258 8

Dedication
This book is dedicated to John McKenzie, Pete McKeown, Pat Farey, Roger Buss and John Archer, all friends and fellow shooters who have sadly passed away during the course of writing.

Photography
All photographs taken by Bruce and Jake Potts.

Disclaimer
The author and the publisher do not accept any responsibility in any manner whatsoever for any error or omission, or any loss, damage, injury, adverse outcome, or liability of any kind incurred as a result of the use of any of the information contained in this book, or reliance upon it. If in doubt about any aspect of custom rifles, or any subject covered in this book, readers are advised to seek professional advice.

Designed and typeset by Guy Croton Publishing Services, Tonbridge, Kent

Printed and bound in Malaysia by Times Offset (M) Sdn Bhd

Contents

Introduction

Ever since man picked up a stone and used it as a weapon, the object has been customized in some way, such as rounding off the edges to make it feel better in the hand or scratching a symbol on it to denote ownership or religious belief. There is an almost inbuilt desire to improve and perpetuate one's possessions. The weapon is no longer simply a tool for getting the job done but is now cherished and lovingly looked after, and often exhibited to show one's status in the tribe. How much customizing is up to its owner and the level of adornment usually depends more on the skill or wealth available to achieve the desired outcome. But the desire is the same: whether it be a weapon, knife, car, house or clothes, it always feels good to have something different that no one else owns. That's the crux of the custom gun ethos, whether it has been built from raw materials to a fully customized state or is limited to a new barrel or a custom stock design. It does not matter: the point is, it's yours.

Many routes can be taken when following the custom rifle project, which often involves trawling through reams of information to select grades of wood, barrel profiles and action designs. The pleasure can be tempered by the long lead times for work to be completed and, more immediately, a huge dent in your pocket. Regardless of the size of the project, however, it will all be worth it as pride of ownership of

Ownership of a custom rifle is not simply about enjoying the look and feel of the rifle. It has to perform perfectly in the field where it belongs.

The author, aged twenty one, with his first custom rifle, a Harry Lawson Cochise thumbhole stocked Sako with Shilen 5.5 profiled varmint barrel in .22-250 cartridge.

the completed rifle as a unique item has its own rewards. Modern day firearms have evolved to such a state that price and quality go hand in hand and the great majority would find that a factory rifle met all their hunting or target shooting needs. There may be a time, however, when doubts start to form and the 'what if' questions come to mind. A Ford Focus, for example, will get you from A to B, but why not make it stand out with a special paint job, magnesium wheels, or perhaps a chipped engine and tuned exhaust note? Life would be boring if we all went around in the same old cars. Guns are exactly the same, but enthusiasts mostly go about individualistic in a more sedate fashion.

I was that young chap who bought a standard rifle off the shelf, though it was actually second-hand because I could not afford any better. It didn't, however, stop me tinkering with it to improve the balance, looks and accuracy.

For more than three decades I have had the good fortunate to enjoy shooting and writing articles for three British shooting magazines, testing and also owning some fantastic custom rifles. I want to share the process of what to look for when you decide to follow this route. Once you have started it is hard to turn back, but the rewards are worth it.

First you have to look at the whole picture, how much you want to spend and where that money may best be used. Often all that it needs is a new custom barrel to pep up a tired old rifle, but these have shot up in price, especially if a wildcat calibre is ordered. It is not only the raw materials such as the barrel itself, but you need to choose a competent riflesmith to undertake the work. The ancillary kit, such as reloading dies, case forming dies and custom-ordered chamber reamers, add to the expense. You should also consider the overall time of construction and the higher degree of skill involved. Not everything, though, may go to plan. You have to think of a rifle as a finely tuned instrument: all the parts must play in harmony together to achieve the perfect ensemble. Often you can replace one part only to find it has a negative effect on the rifle's performance, so you need to change another to get it back on song.

Replacing the stock can instantly transform the rifle's appearance into that of a custom item. This can also be the cheapest option if an aftermarket synthetic stock or laminate is used. These are often a 'drop-in' item, so bedding is not necessarily needed and this is something you can achieve yourself. But beauty can be skin deep as rifles are fickle mistresses. Unless the unity between the barrel and action in the stock and bedding is perfect all that lovely new wood is of no benefit.

Custom rifles often grow with the owner. Unless you are careful, what started as a re-barrel job can soon involve a custom stock, new trigger and fancy chequering. Even then sensible choices are necessary to achieve the perfect rifle, which is what this book is about. Chapter by chapter we will look at the individual parts of the custom rifle, laying out the choices and pitfalls you need to be aware of before you embark on a journey towards your dream rifle.

There are three major types of custom rifle:

Factory custom. A special edition or limited run of a particular rifle issued with upgraded finish, engraving or select woodwork.

Semi-custom. Probably the most popular and cost-effective way to obtain a custom rifle. Here an existing factory rifle can be 'customized' by adding aftermarket upgraded parts, such as Match grade trigger units, Select Match grade barrels, blueprinted actions and

(Top) Factory uprated Schultz and Larsen in 300 Blackout; (middle) Lynx semi-custom; (bottom) Norman Clark full custom .35 Whelen AK rifle.

high-grade wood or synthetic stocks. There are also what may be called 'semi-custom' designs, where a rifle manufacturer , such as RPA, Kimber or Lynx, produces a rifle of upgraded quality that can be ordered to a customer's requirements.

Full-blown custom. As the name suggests this is a rifle designed from the floor up exactly to the customer's requirements, utilizing the very best precision parts and highest quality to produce a one-off rifle.

There are many excellent custom riflesmiths in Britain and as a nation we have an incredible depth of knowledge and talent at our disposal. Custom rifles have changed a lot in the last century, from refashioned military rifles taken into the game fields to fully blown, computer-designed and engineered modern marvels. The skill remains the same, but the design and individual tastes have changed across the generations.

Classic rifles that use the ubiquitous military Mauser 98 action are still perhaps the epitome of a true British custom rifle, with meticulous attention to detail and long man-hours required to build what can only be described as pure art. These Mauser actions have spawned many of the world's classic rifles: old Argentine or Oberndorf Mausers are much sought after as the basis of a purely custom creation. We will look at many classic-styled custom rifles made in the old ways and some with a modern twist, but today's custom rifles are more accessible due to the vast range of aftermarket products, available at reasonable prices, that can be custom combined to achieve your heart's desire. Again we will look at each item used in the make-up of a custom rifle – action, barrel, stock, trigger, scope mount and accessory choice – and delve into the world of wildcat cartridges and how cartridge choice and precision reloading equipment can extract every last ounce of performance. You may, though, reach the stage in your shooting career when the standard factory rifle just does not satisfy your desires as your tastes evolve alongside a maturing interest in rifles.

In the past most custom rifles centred around the classic English design. This incorporated English or French walnut stocks with vivid figuring, subtle hues, long raked profiles and a shadow-lined cheekpiece with exquisitely executed chequering. Add to this the flawless, deep rich blued metal finish, with perhaps a custom set of scope mounts, and you have a very British custom item admired around the globe. What factory rifles can never achieve is that close tolerance of manufacture that custom rifle builders can obtain. Where the factory rifle ends, the custom rifle begins. A custom rifle builder, for example, is able not only to enhance the look by sculpting the metal but to also apply a variety of finishes such as Teflon, classic blue, Duracoat or glass peened. In my opinion, however, custom perfection lies under the skin as a precision riflesmith has the knowledge to achieve a perfect fit for better handling and accuracy.

Whereas a factory rifle's trigger may be heavy or have creep in the travel, or it exhibits a bedding area that is sloppy and bolt movement is stiff, a good custom riflesmith can reprofile the bolt and hone the finish to ensure that there is total contact on the lugs. The trigger can be replaced for a benchrest quality item, or honed to the perfect trigger setting and the action bedded into the woodwork or synthetic stock option so that it fits like a glove, using the finest bedding compound and aluminium pillars to achieve stability between the metal and woodwork for consistent accuracy. There are also a myriad of small details that contribute to a rifle's overall finish and fit. These may only be apparent to the skilled eye of riflesmiths themselves, but they become manifest in the overall handling of the finished rifle.

Some people seek pure accuracy in a custom rifle and will demand perfection in this way without worrying about the looks. Here deep bluing and exotic woods count for nothing and all the custom work goes into achieving perfection in the fine tolerances between metalwork and an exact barrel fit and concentric chambering. Work like this is not as visible as a lovely walnut stock, but it is the basis of nearly all the finest British custom rifles.

The wood used for the stock starts as an uncut blank. You should start by selecting the quality or grade, ranging from plain, fancy

Classic designs such as this very English TT Proctor stalking rifle never go out of fashion.

and extra fancy to Exhibition grades and selected one-offs. You then need to choose the type of wood. Should you consider classic walnut, such as the English or French varieties that are becoming harder to find? Or perhaps it should be an American Claro, Bastogne or Turkish walnut? Many people at the custom stage opt for an untypical wood, such as maple, with bird's eye or tiger-striped varieties remaining popular. Exotic woods like cherry, Zebra or Hyedua also make attractive and unusual custom stocks. Many stocks will cost a good deal more than a complete factory rifle and are becoming increasingly difficult to find, hence the trend towards alternative woods and laminates. But don't forget the synthetic route. In Britain there is massive interest in long-range shooting and deer stalking, and many shooters believe a custom rifle must be fitted with a synthetic stock for durability and stability in all weathers where a wood stock might warp at the wrong moment. A synthetic stocked rifle is dependable out in the field and can be made from fibreglass, plastic, nylon or Kevlar; a walnut stock is reserved for more traditional sporting arms. Many synthetic stocks are 'drop-in' items, inletted at the factory for your rifle's action, but all would benefit from custom bedding in the search for perfection. Laminate stocks are also very desirable and span the divide, combining features of a traditional wood stock with the stability of a synthetic. As the name suggests, these are made from thin layers or laminates of wood glued together with an epoxy compound for stability and rigidity. The chosen wood can be traditional walnut, but far more common is cheaper birch wood, often coloured to give a distinctive pattern when cut and polished. Joe West Custom Stocks has a superb array of laminate stock types, colours and styles for any semi- or full-blown custom job, while off-the-shelf GRS laminates offer a radical design to enhance any custom project. For a handmade custom stock that is really one of a kind, however, you could look consider something like the Steve Bowers thumbhole stock for sporter, varmint or F Class custom rifles.

A stock can be adorned with many a custom feature. With walnut the finish is everything: usually a classic hand-rubbed oil is desired to contrast with superb hand-cut chequering with

Steve Bowers is one of the new breed of custom rifle builders whose exquisitely beautiful handiwork is capable of extreme accuracy.

a fine line configuration, or there might be an elaborate pattern, such as basket-weave or fish scale. Adding exotic woods to the forend tip and pistol grip is always popular, as are custom butt pads of rubber or metal and floor plates for concealed ammunition storage.

Actions and barrels are key for any custom rifle and remachining an existing factory action is cost-effective. Actions such as the Mauser 98, Remington 700 and old Sako or Tikka models have been long favourites for custom projects. These can be tweaked with a process called 'Blueprinting', in which actions are machined concentrically around their axis and the internals similarly machined or trued up. This ensures a perfect fit, stress free and a uniform platform for the stock and internal action workings. If the bolt is then sleeved by adding a precision fit collar and truing the bolt lugs for a smooth, precise lock-up, you start to see the route a true custom rifle can take. Many, though, want a new custom-built action and there is now a huge range of custom-built designs just waiting for your new rifle. All sizes, shapes and models are available. These can be single shots or magazine fed, Mauser 98 clones in stainless steel or pure benchrest models. All have a common theme: they are made to incredibly tight tolerances, far more precise than any factory action, and thus give any custom rifle that sense of class and a head start in their accuracy over a basic factory unit or even a blueprinted one. The action is only as good as the barrel and in many regards this is the pivotal feature of any custom rifle that you want to shoot accurately and not just look good.

The barrel, as with the other components of a custom rifle, has many variations and can be a minefield for the uninitiated. You first need to choose whether to go traditional with a blued finish or maybe rust blue? Should that be a matt or highly polished finish? Or do you go for a stainless steel barrel for longevity and all-weather attributes. Again this can be matt, polished or even coated with Teflon, paint or tough military specification coatings such as Cerakote. A rifle's true beauty, however, is all on the inside and not necessarily obvious to a potential buyer. That's what you are buying: a precision cut or button rifled barrel with a honed surface that runs true along its entire

Custom means just that: you can have exactly what you want in a rifle. This beautiful .22 BR custom rifle is from Valkyrie Rifles.

The Venom Arms Company in Birmingham was the premier custom house. Ivan Hancock and Dave Pope made the best custom air rifles in the world.

length. The chambering to a calibre of your choice needs to be similarly concentric to the bore so the bullet runs perfectly true when fired. Again there is another headache as you need to specify chamber dimensions such as neck diameters, leade or throat length to suit varying bullets, rifling twist and number of grooves, not to mention muzzle threads, muzzle crowns and muzzle brakes.

The real beauty of any custom rifle is that you have a chance to order exactly what you want. Take the chance to build the rifle to fit you, the owner. Face the challenge and buy the best! These are just the most basic considerations of any custom rifle build. Throughout this book we will look at specific parts of the rifle and show the choices that can be made, the pitfalls and advantages of certain items, the finishes available, how your new creation can be fed with custom loads and, of course, all those little extras that make a rifle very personal.

THE LURE OF THE CUSTOM RIFLE

My journey on the custom rifle trail started with air rifles and Venom Arms was instrumental in fostering my desire for and love of all things custom. Ivan Hancock and Dave Pope were masters of their game and turned ordinary airguns into works of art. I remember drooling over the pages of *Airgun World* and *Air Gunner* magazines to see their latest creations. In time, as money allowed, I too became the proud owner of Venom Arms custom air rifles, as well as firearms from their Custom Shop. My HW77 Hunter in .22 cal with custom Walnut Hornet stock and .20 Sidewinder Tomahawk are still used regularly and shoot as well as when they were first made nearly twenty years ago, bringing many rabbits and pigeons to the Potts household table.

Custom rifle owners are often a bit quirky – it goes with the territory. I don't take myself very seriously and often there are references to *Monty Python*, *Father Ted* or *Fawlty Towers* in my articles and custom rifle themes. The time when I dressed in a *Where Eagles Dare* camouflage jacket to shoot a muntjac deer with a custom German K98 sniper rifle, however, might have been taken a little out of context.

It all started when I ordered my first custom rifle after an inheritance back in my twenties. I minutely examined all the magazines detailing all I wanted my rifle to have. In the end, due to the need for precise benchrest quality actions and barrels, I had to look to America where about 90 per cent of the precision items then had to be sourced. I settled on a .22-250 cartridge with a tight neck chamber in a Shilen Match grade, stainless steel number 5.5 contour barrel with an 11-degree muzzle crown. This was precision fitted to a new Sako A2 medium action by Shilen themselves. and then all the metalwork was matt blued for the action and glass peened for the barrel. This was then married to a extra fancy Claro walnut stock by Harry Lawson in his classic thumbhole design called a Cochise. This later became my favourite stock design, as will be apparent throughout this book. The total cost in those days was £1275, a fortune to a twenty-year-old as a new factory rifle could be had for £300. Nowadays, however, £1275 will buy only an action, and even then the choice is limited. Sadly I had to sell the rifle to fund my photographic business, but fate can spring surprises and I managed to buy back this rifle back some twenty-five years later.

Sometimes custom rifles can take you down strange paths of desire and often there will be a theme to a single rifle or collection that has some particular resonance with their owners. I had my own epiphany with a series of four custom rifles that stemmed from an initial concept I had been mulling over for a while. It neatly illustrates the route that a would-be custom owner will take and how different rifles, although similar, can be very different. The theme was 'The Four Horsemen of the Apocalypse', inspired by the New Testament book of Revelation 6:1–8. The four horsemen were sent by God to wreak havoc on the impenitent havoc on earth. It seemed a good idea to have a set of four custom rifles that would

The rifle that started it all: my Sako custom A2 actioned .22-250 tight-necked chambered Shilen stainless steel barrel bedded into a Harry Lawson Cochise walnut stock.

The Four Horsemen of the Apocalypse (top to bottom): Plague, Famine, War and Death. As their names imply, they have been very effective custom rifles in the field.

similarly diminish the vermin population down on the farm, here on earth.

The first stage came when Dave Tooley, a US gunsmith, built me a custom 6mm PPC rifle in 2001. This was the best looking and shooting rifle I could afford at the time and was dubbed the 'Plague rifle' by all who saw it in action due to the unnatural way it laid waste to the hooded crows that my parents found were such pests in Scotland. The name stuck. Later, in conversation with custom rifle-maker Mike Norris from Brock and Norris Custom Rifles, we agreed it would be fun to have four rifles, each named after one of the horsemen, that followed the general appearance of the original Plague rifle. I looked up the original biblical reference and found that the four each rode a horse of a different colour: Famine had a black horse, the red horse was for War and Death rode a pale horse. The first to appear, on a white horse, was Pestilence or Plague. Some translations prefer to call this figure Conquest, but that would spoil the story, so Plague it was.

An unholy quest begins: Plague

A quick description of the Plague will give some idea of the format of each rifle and the thought processes that go into planning a custom rifle (or four).

I like my long-range varminting and the look of a heavy-barrelled varmint-style rifle, especially if it is shod with a thumbhole stock and better still if it has a custom action, especially if a wildcat calibre is chambered. The 6mm PPC custom Dave Tooley rifle sported a Hart barrel with 1 in 14in twist rate for the lighter weight bullets with six flutes, and it was threaded for a $^5/_8 \times$ 18in pitch sound moderator. This was fitted to a McMillan MCRT custom action that had a semi-cone bolt system with only an extractor and non-ejector, so I could pick out the brass without it falling in the mud. This gave me the best possible chance of precision. Fitting a Jewell trigger unit set at 10oz weight transforms any custom product into a tack driving machine. All the metalwork was then clad in Teflon coating to achieve a muted finish, perfect for hunting as it

The Plague rifle in 6mm PPC is superbly accurate. It has a Hart barrel and McMillan MCRT single-shot action, all bedded into a black Lazzeroni stock.

The single-shot McMillan MCRT action does not have a magazine cut-out and is very solid. Its Teflon coating means that its operation is as smooth as butter.

eliminated any shiny surfaces that might spook the game. The best bit was the stock. I love the stocks from Harry Lawson, a stock maker in Tucson, Arizona. My first custom rifle had a Harry Lawson walnut stock, but this time I wanted a synthetic stock and luckily McMillan produced a copy of his design called the Lazzeroni thumbhole stock, which was perfect for my needs. The stock is the single part of the rifle that transforms the overall look and feel of any custom rifle and it will be your biggest choice. The thumbhole just looks right and I really like the pistol grip, which has an offset of some 20 degrees to give a more natural hold. The action was aluminium pillar bedded in Devcon synthetic material and the rifle was topped off with a set of Talley rings and bases to hold a NightForce NXS 3.5-15 × 50mm scope, which it still wears. Finally a superb MAE muzzle can is fitted to reduce the muzzle report to a whisper. Stage one complete: the Plague rifle shoots a 65 grain V-Max bullet at 3087 fps with a payload of 26 grains Vit N133. Nothing within 500 yards is safe.

War

The red horse or War was next. This was to be a fully customized piece featuring a wildcat calibre. The first requirement was the same Lazzeroni stock, so I ordered a black textured Kevlar reinforced stock from Jackson Rifles, ready inletted for the Remington M700 action. I could have used a Remington action, trued or blueprinted, but what's the point when by the time you have paid for that work you might as well

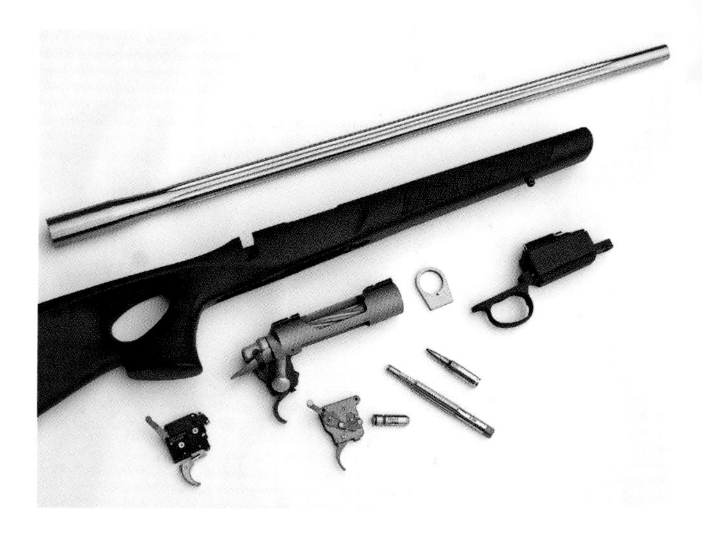

The components of the War rifle, a necked up 6.5 × 47L case to .30 cal custom-made by Steve Bowers to my design. This is the rifle I choose for everyday deer stalking.

The War rifle is superbly accurate and offers great handling in all weather conditions.

Nosler Ballistic Tip bullets with 125 or 150 grains, depending on the size of deer, are very accurate and effective.

buy a true custom action that fits into the Remington bedding platform. I ordered an excellent Predator action from Stiller Precision. The Predator was again synthetically bedded to the stock and this time I fitted a Timney trigger as I like its wider trigger blade and generally solid feel. I also fancied a change.

I wanted a detachable magazine as this was to be a deer gun and this is a handy feature, especially if you need to change bullet weights for a chance at foxes. The HS Precision unit from Viking Arms, with its stainless steel construction and black finish, suited the rifle as well as my criteria and the single final feed of the cartridge is actually very positive. I could have fitted a one-piece scope rail with a 20 MOA (Minute of Angle) bias, but this was to be a deer gun up to 300 yards so two- piece Leupold steel bases were fitted so I could use any Weaver type rings I liked. In fact I choose a set of Apel mounts with the upper section made to fit a Zeiss rail mount system, which here was a Zeiss 7 × 50 T scope with a simple 4a type reticule, thin and with the option of an illuminated central dot. This would be perfect as a no-nonsense, hard-wearing, take any abuse deer rifle.

Now the problem was choosing the right calibre. Having gone this far a standard .308, although very good, would be a bit boring. I stuck to a .30 calibre, but for the War rifle I wanted a small but superbly efficient wildcat design. I thought it would be interesting to neck up a 6.5 × 47mm to .30 calibre. The resulting 30-47L is a simple wildcat to make, superbly efficient and deadly from muntjac all the way up to red stags. It shoots a 125 grain Nosler Ballistic Tip bullet with a load of 41 grains of Vit N130, producing more than 3000 fps velocity at 3026 fps and 2541 ft/lb energy. The barrel was a Walther stainless steel unit with a 1 in 10in twist rate, fluted to reduce weight and a short length of 20in, so that fitting a sound moderator would not make it long overall.

Famine

Famine, on his black horse, represents a very unusual rifle indeed. It satisfies my odd obsession with big bullets travelling very slowly with a meagre diet of less than 10 grains powder – hence Famine. Subsonic full-bore rifles are of interest due to their ability to send a 240 grain projectile, in the case of the .308 Winchester, at subsonic velocities (below the speed of sound). What you have is a 500 ft/lb energy rifle that is no louder than a .22 rim-

The Famine rifle is a silent killer, designed to shoot subsonic ammunition with a built-in sound moderator. It has been restocked by Ivan Hancock of Venom Arms with a Cochise walnut stock.

fire subsonic gun when fitted with a sound moderator, although it is no good for anything other than playing, military use and foxes (in my case). It does not matter as I often set out a course of fire across a few fields in Scotland up to 500 yards with steel silhouettes adopting a ten rounds to spot, range and 'one round, one hit' scenario, which really makes you think about the shot. The bullets drop like a stone, but energy loss is minimal even at 500 yards and the wind has little effect as it travels below the super and transonic air turbulence. A 200 grain Lapua B416 bullet, for example, propelled by only 9.25 grains of Vit N320 powder achieves consistent subsonic velocities. I have used other weights and powders, such as IMR Trial Boss and Tin Star N350, but the Vit N320 works in my rifle.

This rifle is an original Sako Factory SSR (Sniper silenced rifle or Super silenced rifle), built on a high quality Sako A2 action and fitted with a short 16in barrel with a 1 in 10in rifling twist rate. Surrounding this is a 2in sound moderator shroud that fits back to the action and extends 12in past the muzzle, where a series of baffles are sited. Using subsonic reloaded ammunition, the rifle just coughs: you hear the action tighten up and the firing pin strike is actually audible. To make it into my Apocalypse set I had Ivan Hancock from Venom Arms re-stock the SSR in a Lawson walnut thumbhole stock. Ivan sourced this from his friend Trooper in the States and crafted it into a totally silent rifle with superb handling.

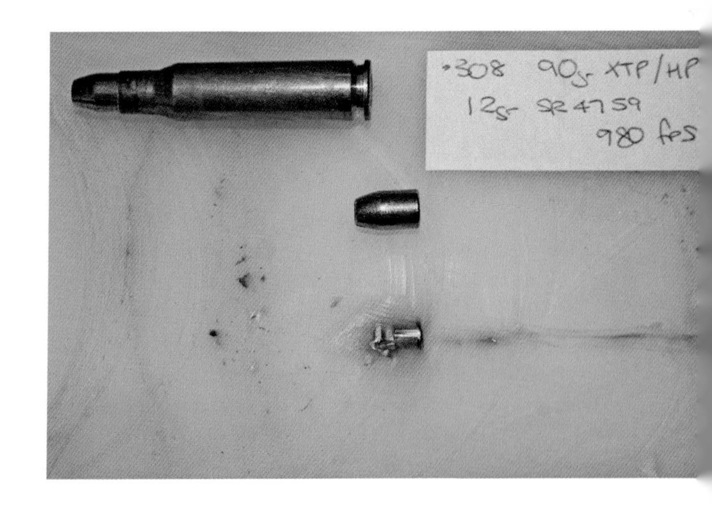

When using my .308 Win Famine rifle I find that the best load for effective subsonic stopping power is the Hornady 90gr XTP bullet and 12.0 grains of SR4759 powder.

The Death rifle is a Sako .20 BR rifle made by Venom Arms with a Pac-Nor barrel and specially selected Claro walnut stock.

Death

Death rides a pale horse, which is appropriate for the final rifle in the quartet of misery, the good-looking Venom Arms .20BR rifle. This time Ivan Hancock and Dave Pope's son Steve sourced a select, not quite Exhibition grade, Claro walnut stock called a Cochise from Harry Lawson. He inletted the forend with a diamond tiger maple inset and inletted the stock for a Sako 591 action sourced from Gregor at R. McLeod's of Tain. Old Sako actions make a great donor action for custom projects. The bolt was jewelled and the trigger was a factory-set trigger unit. The barrel was a 26in Pac-Nor stainless steel fluted masterpiece of .204 calibre with a 1 in 12in twist rate. Pacific Tool and Gauge made me a reamer and head space gauge for the .20 BR case machined without a throat or leade because the .20 BR cartridges can be hard on barrels. Up front is an excellent MAE 32mm muzzle can for a very quiet report.

I use a load of 30 grains RL15 powder and a 40 grain V-Max bullet for 3885 fps and 1341 ft/lb energy. Alternatively for real speed, a 32 grain Sierra Blitz King reach more than 4300 fps with just 31 grains of Vit N133. The .20 BR is topped off with my favourite Varmint 4-16 × 50 scope from Schmidt and Bender with side parallax and dot reticule in the first focal plane. The multi-dot reticule is fine enough even at the highest magnification and mimics the .20 BR's trajectory well out to 750 yards. The .20 BR is a great cartridge and I will probably have it rebarrelled when it gives up the ghost with the faster 1 in 8in twist barrel to handle the heavy 55 grain Berger bullet that I use in my .20 Satan wildcat.

Conclusions

Therapy may be the only course of action here but it shows what can be done with a little thought, yielding a rifle collection that

One of my favourite crow cartridges is the .20 BR cartridge with 40 grain Hornady V-Max bullets that reach more than 4000 fps.

The Twenty calibres are currently very successful in Britain, ranging from the .20 Hornet up to the .20-250 (right). The best balanced is the .20 Tactical (third from left).

gives hours of varied shooting pleasure. This series of rifles shows that every aspect of design and component choice can be varied around a common theme to achieve a totally different rifle for differing purposes, yet still please the owner, which is what a good custom rifle is all about. Choice of custom items is often made on availability. For a long time most custom items came from the USA, but truly bespoke items from home-grown custom gunsmiths can always be fashioned and quality actions, barrels, stocks and triggers are now made to rival even the best from the US. Unfortunately it soon becomes an addiction. You are always looking for the next best thing, a strange custom item or a weird calibre to satisfy your craving. Don't blame me if your wife finds you mumbling to yourself about polygonal grooved fast barrels and case hardened Talley mounts.

Chapter 1
The Stock

The stock is often the most striking and distinctive part of any custom project and instantly transforms an ugly duckling of a rifle into a swan. Be it traditional wood, newer laminate materials or the practical synthetic, a visual and ergonomic stock design really transforms a rifle. Like every part of a custom rifle, however, beauty cannot just be skin deep. As much attention must be given to the interior fit as to the exterior finish. There is a hugely enjoyable range of choices in design, materials and exterior finishes from which to choose, but with a custom rifle the bedding of the action and barrel to the stock is vitally important for accuracy. The stock should be considered an extension of the rest of the rifle.

WOOD, LAMINATE OR SYNTHETIC

One of the major decisions is whether to go for a wood, laminate or synthetic stock. Traditionally walnut has been the choice of the rifle stock world, but many other types of wood have been used as necessity has dictated. This has led to a diverse variety of wood species finding their way onto the world's custom rifles. Some are exquisite works of art and others are a reflection of the individual's preferences. A gun stock needs to be strong but not too hard, as otherwise it may crack or shake, which can be a problem with oak. It must be able to take a chequering tool for ornate chequering at the finishing stage. A standard walnut will take 16 lines per inch

Shooters can choose between laminate, wood or synthetic stock materials.

One of the enjoyable aspects of selecting a wood stock for your custom rifle is that you can choose the type and grade of wood. Here are some well-figured and seasoned walnut blanks.

for chequering, while a high-grade, denser walnut will be able to hold 24 to 26 lines per inch for ultra-fine, delicate chequering. Selecting the correct tree and buying sufficient material from the one source makes sense for stock makers as it ensures a uniformity of supply and quality, but this is not always possible. It also ensures that only properly seasoned material is used, which is the most important aspect of a stock.

A freshly sawn tree has a very high moisture content of 30 per cent, depending on the species, but this figure has to be brought down to well below 10 per cent before it is any good for rifle stocks. Unless it is properly seasoned there will be shrinkage, warping or cracks, making even the best wood unusable. Oak is not a good gunstock as it dries and cracks, or shakes too much despite being strong. That's why walnut and maple have become popular. By the same token you want it strong enough to withstand recoil, so a close grain structure, but with enough good colour and figuring, is still needed to add appeal. That's why hard woods like poplar and birch are used in laminate stocks, not for their beauty but for their strength as the grain structure is close and dense. It's a dilemma but wood has been refined and cross-grown over the centuries to achieve some extraordinary custom stocks. Unfortunately good quality walnut is getting harder to find as a walnut tree can take between fifty and seventy-five years to grow big enough to be cut for timber. Older stocks of trees have been exhausted in traditional growing regions, such as England and France, and now a Turkish and American hybrid walnut seems to be yielding better quality wood.

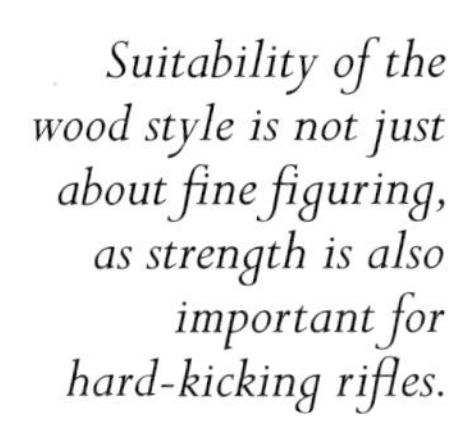

Suitability of the wood style is not just about fine figuring, as strength is also important for hard-kicking rifles.

Characteristics of wood for custom stocks

- Straight, dense and close grain.
- Dries to a consistent state with a stable moisture content.
- Minimal degree of shrinkage, cracking and splitting.
- The wood must be machine ready and capable of being carved or inletted to a precise degree.
- If it is to take a stain or finish when completed it should not be too oily or dry.
- How well is the wood figured for a particular task, with good figuring in the butt section but closer grained through the pistol grip for strength.

Above all, every stock and application is different. Every project deserves a considered choice.

HARVESTING THE WOOD

Choosing a good walnut blank for a gun stock is a tricky decision dependent on how the wood blank is harvested from the walnut tree's trunk. The figuring on a good walnut blank is made from mineral lines striating throughout the wood blank. The degree, density, curvature, colour and pattern is different in each individual trunk, but certain cuts through the trunk yield different patterns. These mineral lines travel up and down the end grains of the blank and their formation is the basis of nearly all figuring to the custom stock.

Three main cuts are available to achieve the differing grain patterns and quality figuring for which the walnut tree trunk is renowned: flat sawn, three-quarter sawn and quarter sawn. Quarter-sawn blanks are plainer and more regular with mineral lines and colour that is even and duplicated on both sides of the blank. They are cut with one outer edge pointing at the bark and the other towards the centre of the trunk. Three-quarter-sawn blanks have more variation in the mineral lines, with swirling, curvature and colour variations to the grain that give individuality to each blank as it is cut. This variety comes from whether the sawn blanks are taken from the top or bottom of the trunk, with horizontal cuts across the trunk to the centre line. Flat-sawn blanks are the best quality and come from the inner part of the trunk where the best grain patterns are present. When cut you can see the density of colour, marbling and figure by looking at the end grain, indicating the type of figuring to the sides of the finished stock that will give a superior grade of wood.

The layout of the stock pattern is also vitally important, especially on high-recoiling rifles. Stability across the entire stock is important, but the way the grain pattern flows through the vital areas, such as the pistol grip and bedding areas, is the difference between a usable stock and one that

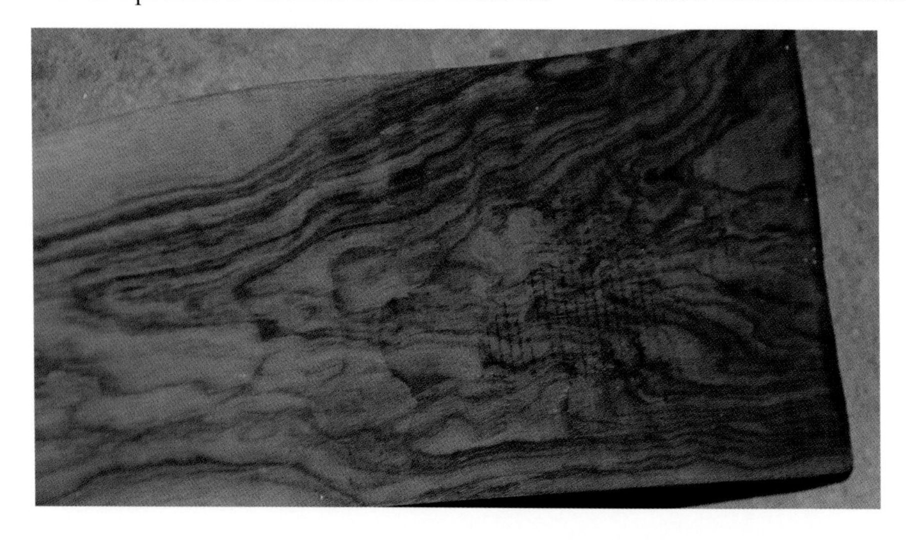

Flat sawn blanks yield the best figuring patterns across the whole wood blank.

Three-quarter sawn blanks have very good figuring running through the stock.

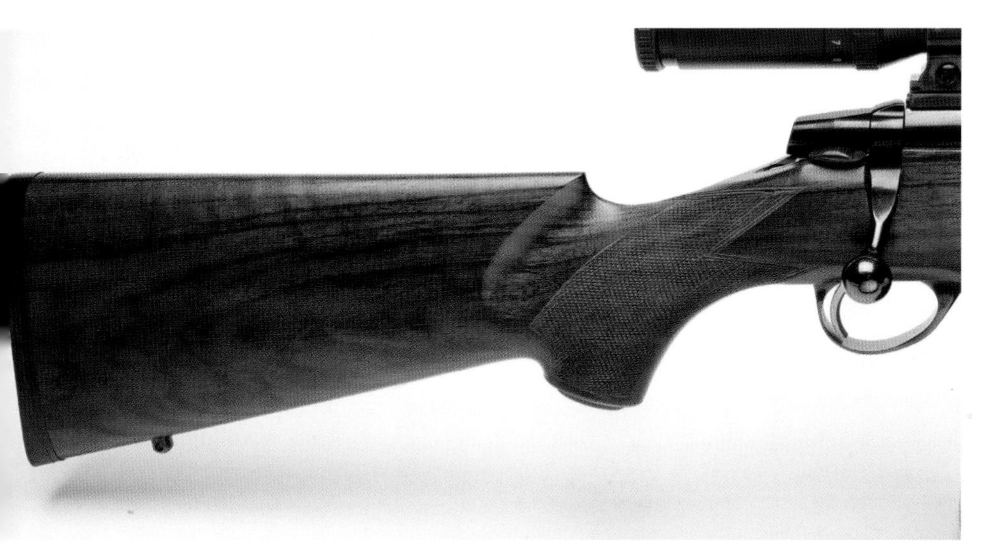

Quarter-sawn blanks offer attractive figuring with more open wood spaces.

The end of a walnut blank showing how the grain runs from one side to the other, depending on how it has been sawn. Check the figuring on both sides before you buy.

will not do. The grain should ideally be straight through the areas that take the torque stresses from firing the rifle. If the grain goes in at an angle, the stock then becomes liable to splitting or cracking. The figuring and colour variations can vary in the pistol grip, forend and bedding area, but the grain direction must be straight. That is why the stock on a big game rifle has a plainer pattern for strength, not beauty.

WOOD TYPES

Walnut

Walnut is the most commonly used and loved wood for custom projects. It has a vivid and varied array of colour, grain, streaking and pattern from dark lustre to sandy brown, with intense swirling or curls. It is also easily worked but hard enough for strength. Although quite open-pored in structure, after sanding it allows a good penetrating oil and finish to be applied and sustained.

English (*Juglans regia*)

The best of the many types of walnut available is usually termed English, but it is also known as Circassian, French or Turkish English. It is defined and distinguished in appearance by the soil and area from which it is derived and where it is grown. This is the most prized variety of walnut as it can vary from light sandy shades to dark rich chocolates, with fine black streak figuring to a dense, swirling marble cake. French regional 'English' walnut, for example, can be sandy with vivid black streak figuring, while trees originally from eastern Europe are better in colder climates and have a slower growing period. This results in a denser and more varied marble cake figuring. Both types, however, can be called English walnut. It can be confusing but prospective clients for a custom stock usually know the style and figure pattern they are after, even if they do not know the real name or origin of that blank.

English walnut is much sought after and good quality blanks are now rare, but it makes a very elegant wood stock for a classic English rifle.

Dark figuring running parallel along the blank makes for a beautifully flowing grain pattern.

Bastogne Walnut

Bastogne walnut is a quickly growing hybrid cross between English walnut and black walnut. It is not commonly used, but is sometimes asked for by shooters who require a dense, hard and strong walnut for rifles with a heavy recoil as it has closed pores and regular grain. With regard to the colour and pattern, Bastogne has a typically shady, almost beige-green tinge, but can also exhibit high figuring and strong a fiddleback, but it varies enormously from tree to tree and region to region.

Black Walnut (*Juglans nigra*)

Black walnut is far more common and usually has the traditional American reddish overall colour to the stock blank. It displays a wide range of figuring patterns with excellent selections of burr, crotch or fiddleback pattern. The quantity and quality available means that this is the first walnut of choice for many since the reasonable price and relatively good figuring and colour are enough to satisfy all but the connoisseur.

Claro Walnut (*Juglans hindsii*)

This is another type of black walnut and is widely used by custom stock makers as it has a very good degree of figuring and colour variation, which are desirable qualities where customers' tastes can vary. Although this walnut is a bit softer, this is not a problem and it is as strong as any other provided the stock is properly bedded with aluminium pillars or bedding compound. One of the reasons Claro is popular with custom rifle makers is that it grows quickly and so larger trees yield more blanks. There is also a better chance of deeper figuring patterns that extend from one side of the blank to the other, usually three inches in width before shaping. Colour variation is also very pleasant, from deep red and purple hues intertwined on a pale or deep yellow background wood, with even greens and dark chocolates adding to great variation.

Black walnut is graded from standard to Exhibition grade. This special selection piece has everything you want in a custom rifle, displaying great figure and colour.

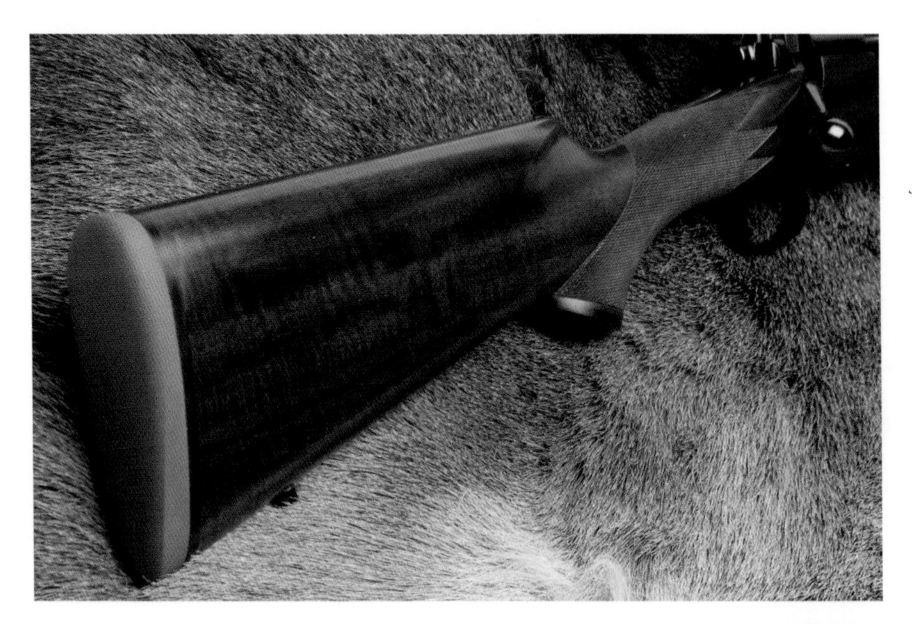

Claro walnut is more readily available and offers a popular blend of straight grain and fiddleback pattern throughout the stock.

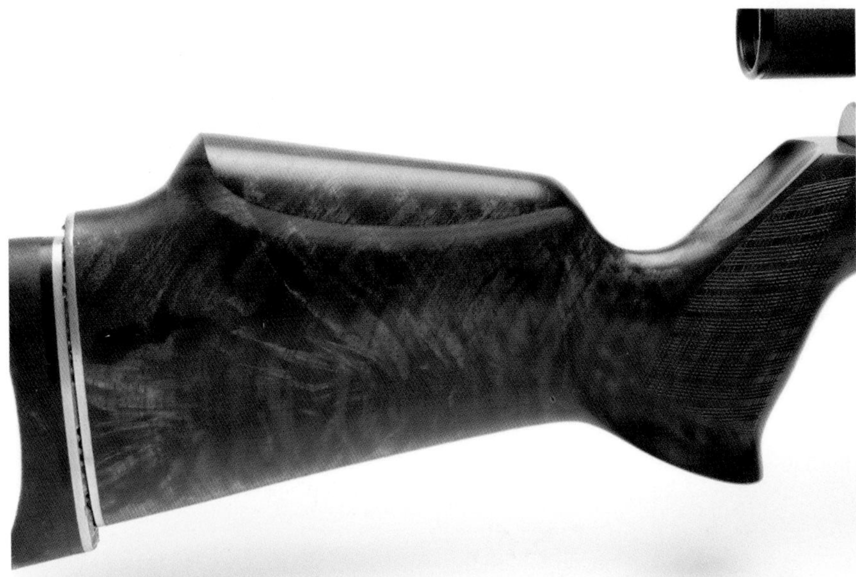

Any shooter would be proud to own the gorgeous Claro walnut stock on this Venom Arms custom .22 air rifle.

Maple (Acer macrophyllum)

Maple is a stock type that divides opinions. Most people like a darker stock colour with plenty of differing patterns, but maple is nearly always light or buff in colour. In my opinion that is its charm since the figuring is more subtle and it can accentuate the metalwork that sits in it. That's not to say that some maple cannot be highly figured; it can be, but with more regularity than walnut. The main types are fiddleback, shell flame and bird's-eye. All have a lighter base wood with more distinguished grain patterns of darker buff to give a very pleasing pattern. Bird's-eye maple, as its name suggests, has numerous small swirls all over the stock arranged in darker rings that resemble the eye of a bird. Shell flame is an attractive variant that is often called quilted maple as the darker patterning resembles a shimmering flame on a light background, giving it a striking three-dimensional appearance.

Maple stocks offer striking effects with a paler colour complemented by darker figuring.

The shell flame patterning on this Maple blank would make a unique custom stock on the right rifle.

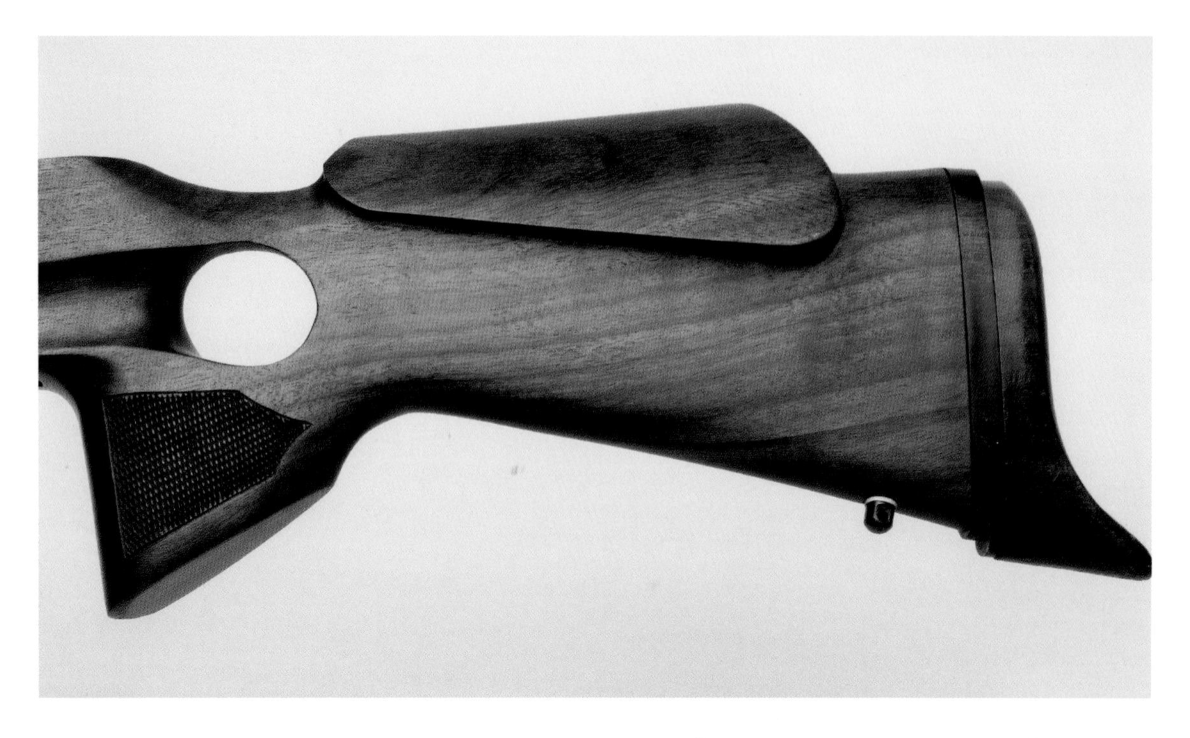

Hyedua wood, as used by Theoben Engineering on their gas ram rifles, has good colour and unusual figuring.

Regardless of the exterior appearance, maple has always been revered for its very dense and strong grain pattern. It makes an ideal gunstock that, due to its hardness, also takes carving and other stock embellishments.

Hyedua

This is an exotic African hardwood that is pronounced *shed-u-ah*. It became very popular in the 1990s as an alternative to walnut, which was becoming increasingly more expensive even as the quality dropped. Theoben Engineering, the inventors of the gas ram air rifle system, used Hyedua to stock their best air rifles, such as the Taunus, Olympus and FAC rated Eliminator. It has a very particular grain pattern and a most striking and attractive honey or golden brown colour, with deep brown grain figuring and often plenty of a fiddleback effect.

GRADING WOOD BLANKS

Which is Best?

This is my favourite part of any custom rifle project where the grade of the wood will make an enormous difference to the overall appearance. You have to really know what you are looking for, however, since it can be a lottery as to what you get if you order a specific grade of one type of walnut strain. Most custom shops have a good selection of wood to choose from so you can see, handle and then decide. Pre-ordering on-line can have its downsides as flaws such as voids, cracks and warps may not be obviously apparent. Also make sure you see both sides of the blank/stock to check that the figuring and colour is the same grade. Be aware of the positioning of your stock shape on the stock blank to make sure of its strength due to the grain pattern and that the figuring flows the right way through the stock. Make sure the fiddleback is vertical and ideally evenly spaced along the entire stock length. The colour should also be the same on both sides and from tip to end.

Everybody has their own idea of what a good wood stock should look like, which is why a custom rifle is such a personal item to own.

A one-off or special walnut stock on a rifle enhances one's pride of ownership. This walnut stock shows lovely figuring and deep dark colour.

Heavily figured walnut looks very odd if the grain does not flow along the lines of the stock design and through the pistol grip. The butt section of the stock is where most attention is focused and the best figuring is on show, as chequering can make up for any shortfalls along the pistol grip or forend. Ideally you want the colour and figure to cover the entire area, making sure there is no sap wood at the edges to spoil the appearance. Also check that the figuring covers both sides well or evenly, otherwise you may have one well-figured side and the other a plain Jane. A good rule of thumb for well-balanced stock figuring is to have the flow of the grain matching the form of the stock profile. I have seen stocks with the figuring facing upwards rather than the more traditional flowing down the butt, and it just looks odd.

The best figuring should be in the butt section as this does not usually have any chequering covering it and so the whole beauty can be displayed. Straight grain or less figuring is then desirable through the pistol grip and forend section for strength. I like to have figuring all over the stock where you can afford it and on small calibre rifles where recoil is not an issue. Remember, however, that warping or twisting due to climate change can be a serious problem.

Grading types used by the major stock makers

Special Selection	**one-of-a-kind pieces**		
Exhibition	100% figuring, excellent colour	AAAA	5
Fancy select	75% figuring, good colour	AAA	4
Fancy	50% figuring, good colour	AA	3
Semi-fancy	10% figuring or better, good colour	A	2
Utility	plain wood	none	1

Grading System

Wood suppliers can use differing scales to describe their wood grades, either using a numbering or lettering system: the higher the value, the better the grade. Often a percentage figure is used to describe coverage of figure and colour to the wood blank.

The grading of stocks varies depending on the source: the four stocks here are graded from standard (bottom) to grade five (top).

Semi-fancy stock gives better than standard colour with some figuring to the stock.

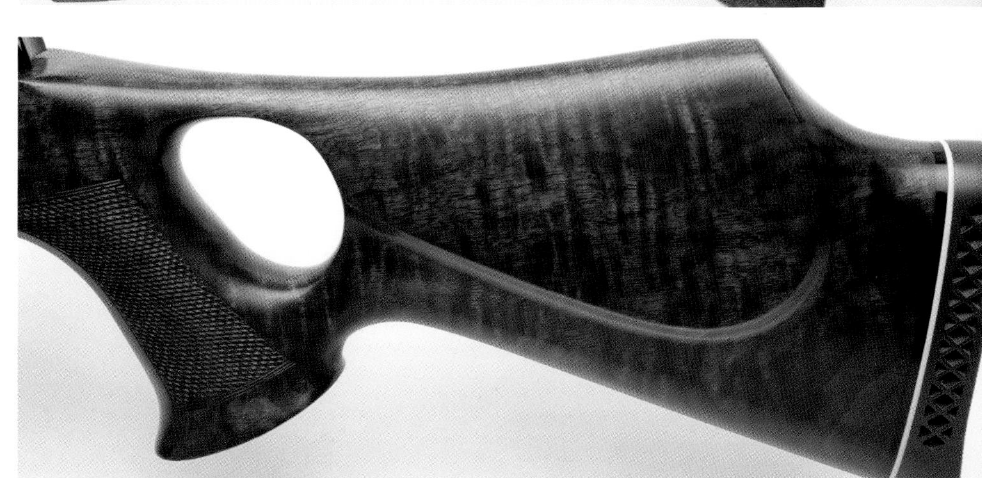

Fancy grade three shows much better colour and figuring pattern on the butt stock and forend.

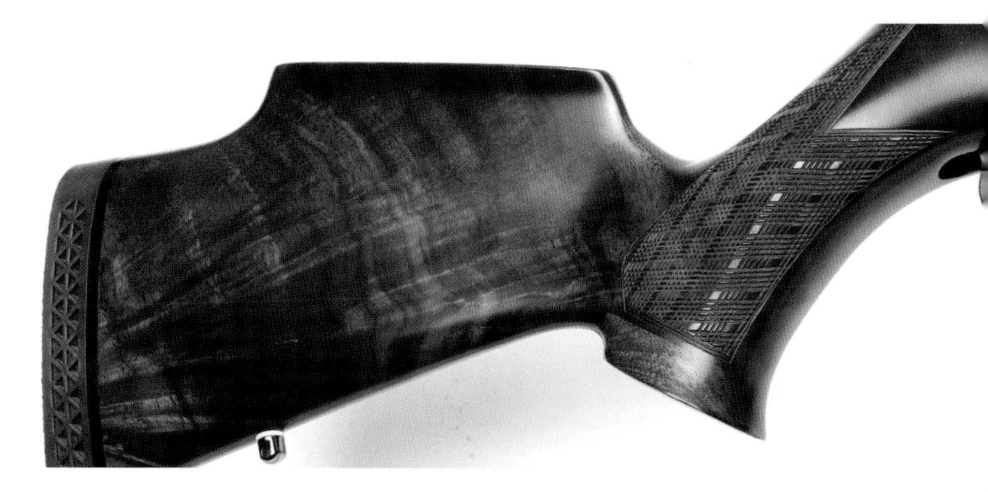

Fancy Select grade four/five shows excellent colour, contrast and figuring.

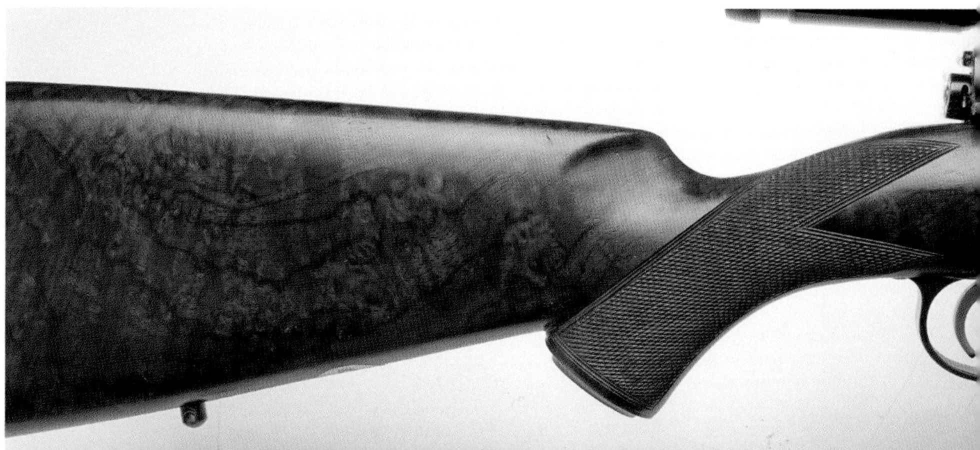

Exhibition grade offers superb colour and figuring.

The interplay between the background walnut colour and darker figuring creates an effect known as marble cake figuring that is much sought after.

Pattern

Any custom rifle is a product of its owner's tastes. Grain pattern and colour are very personal choices. Fiddleback pattern is my personal favourite as this has very striking, near-vertical and parallel lines that are perpendicular to the growth rings that make up the underlying grain pattern. They can be present in most forms of walnut, except English, and really enhance an otherwise plain coloured and grained stock. The fiddleback can be subtle or vivid, but both types share a special quality in that, as the stock is manipulated from side to side in the light, the fiddleback grain shimmers and changes thickness and intensity. That's why fiddleback is very popular on any custom project: on walnut it is striking, while on maple it is often the only grain pattern and really stands out on a plain, pale wood background. Feather and burr walnut pattern is taken from a stock blank closer to the root of the tree and as such has a deeper patterning to the wood and a more sophisticated array of colouring and grain. The interplay between underlying dark grain intersecting with lighter tones of feather crotch that radiate through the stock is very appealing. As the complexity increases so does the price, but you have to be careful that the open layers of growth and patterning are not present in the wrist or action areas, as they are not as strong as a tighter, plainer or finer grain. That is why most of the best feather, burr or crotch figuring is selected on a stock blank for the butt section, while the wood that flows through the wrist and forend can be less figured or fiddlebacked. It still appears beautiful but remains strong.

Marble cake is another term used to describe some of the best figured stocks available. Here the walnut has an amazing combination of colour and intersected figuring that looks, as the name suggests, like the flavourings running through a marble cake mixture. The problem with this type of walnut is that, when it comes to the price, the sky is the limit. Individual blanks that can make the best custom stocks for a rifle could set you back by thousands of pounds.

The Mauser K98 rifle used one of the first laminated stocks. Although chosen mainly for its cheaper production costs, it proved a very strong rifle stock material and modern stocks are essentially the same.

LAMINATES

As the price of quality wood goes up, the use of laminated wood certainly makes financial sense. Not only is laminate cheaper but it also has the benefit of being far more weather resistant than a standard wood stock. This makes it a very practical rifle stock and in some styles it has a charm all of its own. Laminate stocks, as their name suggests, are made from a series of thin wood strips glued together to form a strong multi-layered stock. This gives a very strong stock that is impervious to changing climatic conditions and less likely to warp, as walnut does, and thus become less accurate.

Laminated layers are often cheaper wood species, such as birch or beech, and each layer can be stained a different colour so that they combine to create a layered effect that makes a pleasing replacement for a fine figured walnut stock. These layers are usually $^1/_{16}$in (1.5–1.6 mm) thick. Each layer is colour pressure treated to saturate it and then they are glued together with a tough resin glue and hydraulically pressed to ensure a perfect bond between each layer. As the layers are thin, the colour dye gets right into all the pores and no external water can enter. This makes it highly weather resistant. The glue or epoxy also adds strength, but this is greatly due to the way the materials are arranged. The direction of the wood grain is glued so that it alternates between layers, making a very strong bond. Many makers colour these layers differently, so the resulting patterns can look very good when shaped and sanded, although you never really know what you will get. Some of the first laminated stocks were supplied to the German army during the Second World War for use on their K98 bolt action rifles as the supply of conventional walnut stocks dried up.

Benefits of a Laminate Stock

Laminates ensure a uniform stock material to work with. They allow for a very strong internal and external stock design that is far better than any standard factory wood or plastic stock on offer. Laminates also have the benefit of being highly resistant to climatic changes and are virtually impregnable to water or moisture. They retain their rigidity, which is ideal for maintaining consistent bedding between stock and the barrelled action. Laminates are also very good in conditions at the other end of the weather scale. They do not degrade due to the UV rays from the sun or in response to humidity changes in tropical climes, which makes them a good choice for overseas hunting.

A laminated stock is almost as strong as a synthetic one, but it has the feel and warmth of a wood stock. A laminated structure is also better at deadening recoil and vibrations from the firing process than fibreglass and metal-framed stocks. The only major drawback of laminated stocks, apart from their appearance, is that they weigh more than a similarly shaped wood stock. The extra layers and glue add weight and these rifles are heavier to carry around the woods, but in my view any negatives are far outweighed by the benefits to strength and resistance to warpage.

They are also an economical way of producing large stocks, which are now becoming popular for large-calibre, long-range custom rifles. Extreme-range shooting requires a large and stable platform that can cope with the recoil from large-calibre weapons. Oversized

Starting with a blank, Steve Bowers creates a custom laminated stock by crafting it into a finished product that exactly suits the customer's dimensions for perfect fit and function in the field.

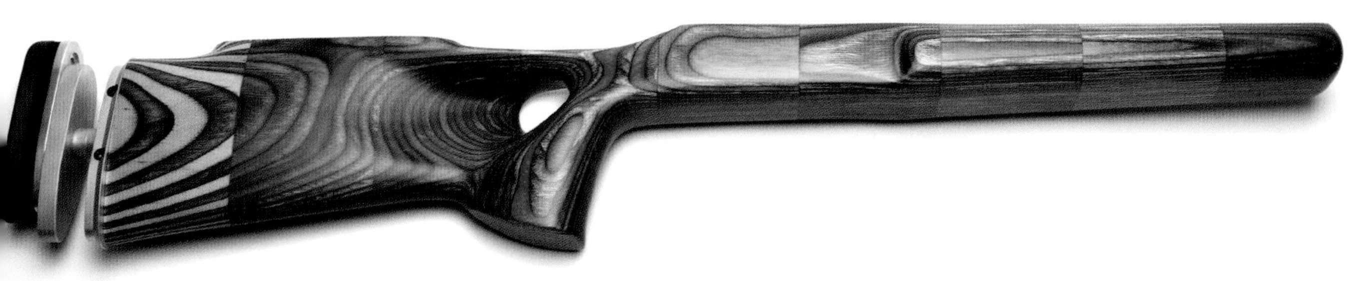

Laminate comes in many colours. Here is a sample JoeWest laminate stock for customers to select their preferred colour combination.

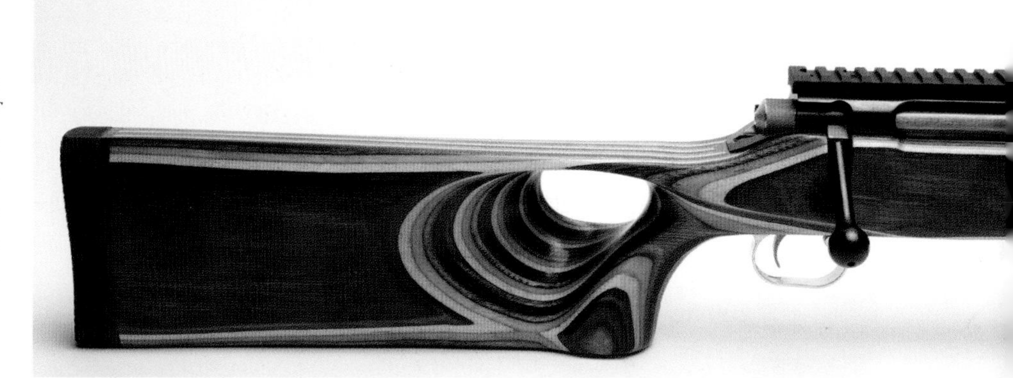

Laminates make strong and weatherproof stocks. They are popular for benchrest rifles, such as this Callum Ferguson bench gun.

A fully adjustable laminated grey stock by Steve Bowers.

forends with wide, flat bases allow the rifle to ride the shooting bags in free recoil situations with the highest accuracy. Laminate can be carved, painted, custom wrapped, or treated with Cerakote, Hydrographics or just a clear lacquer coat. providing plenty of external visual custom features to keep owners happy.

Laminate stocks are among the most popular stock types in Britain, especially for small calibre rimfires as not many synthetic stock makers cater for this market. Therefore the cost-effective nature of a laminate stock and its ability to cope with all weathers makes it a practical and desirable choice on a rimfire rifle that will be used largely as a working gun and not spend most of its life being admired in the gunroom.

Synthetic stocks, as shown here on my 300 RUM Geske actioned rail gun, are very popular due to their strength and ability to hold zero-on accuracy in bad weather. The unlucky crow was downed at more than 1000 yards.

SYNTHETIC

In the late 1960s Chet Brown and Lee Six in the United States pioneered the introduction of a new stock material called fibreglass that had previously been used only on sailing boats. These synthetic stocks created a market for hunting rifles that would be used by real world hunters who demanded a strong stock that would not warp in extreme climates. The stocks were all hand laid with layers of fibreglass and resin glue moulded to form a stock profile. They were a bit rough or utilitarian in appearance, but they were 100 per cent non-zero shifting and nothing about them impaired accuracy. They set the trend towards the synthetic stock revolution as shooters increasingly replaced their wood stocks with synthetic. Brown Precision still makes stocks, whilst Lee Six formed a new company that was later bought out by Kelbly. I have a Lee Six stock on my 300 Rem Ultra Mag rail gun. It is a magnificent stock proportioned for use at extreme range, beyond 1000 yards. It has a moulded synthetic structure enhanced by lead inserts to add weight to deaden recoil, so it can only be shot from a bench. It has a large flat 6in forend with twin stainless steel inserted runners underneath to slide on. This stock design allows free recoil shooting where no part of the rifle touches the shooter and is capable of extreme accuracy at long range.

That was the start of the synthetic rifle stock revolution. In those days wood was the only way to go and anything else was scoffed at by purists. Just look at all the rifles sold today, however, and the majority are synthetic or laminate. Times change. Synthetic stocks are not all the same. As with wood there are many types, some good and some less so. Many synthetic stocks today are just used as they are cheap to produce and portray none of the ethos of a quality synthetic stock you need for a custom project. The term synthetic is also confusing as it means artificial in nature, but the actual composition varies widely. Quality synthetics are built with rigidity, chemical inertness, lightness, strength, noise deadening and water resistance in mind, whereas cheap plastic stocks lack any form of finesse and are often the opposite of these qualities.

Moulded

These are usually cheaper synthetic stocks using a plastic or injection-moulded process. The two halves are then glued together to form the stock design. They resist water but feel hollow, resonate noise and can flex under recoil and handling. I suggest you steer clear. They use a polyurethane foam subtract strengthened by fibreglass but aluminium pins or insets can be inserted to strengthen the key areas, such as the pistol grip, action and forend. This adds weight back to the stock and negates any advantage of a light stock. Moulded stocks that use a solid moulding or two solid halves glued together are better, due to the enhanced strength, but they can be heavy.

Composites

Composites are by far the best custom synthetic stocks. As the name suggests, a composite of materials is used for lightness and strength. They are usually handmade, using an outer mould to form the desired shape of the stock design, into which is laid individual fibreglass cloth glued together to form an outer shell. A polyurethane foam is later poured in to fill the void. This makes a strong yet lightweight stock into which a new custom rifle can be bedded or have a synthetic bedding system installed. Substituting the cloth fibreglass layer for materials such as Kevlar and carbon fibre alters the strength and lightness again but also increases the price. It does mean, though, that a custom rifle can be made extremely strong and rigid for long-range use but still be extra-light for hunting in the mountains. For maximum strength the shell can be filled with an epoxy resin material that fills the void very well and gives a solid feel, yet remains lightweight.

This type of stock has an extended lifespan and is very resistant to knocks and scratches, which is another benefit of any synthetic stock over a quality piece of walnut. Although synthetic these stocks are not cheap due to the hand-built quality of the process but they are now common and very popular. Shooters like the dependable feel of a synthetic stock, the reliability and the accuracy, as there is no need to change zero due to warpage caused by water ingress.

Composite synthetics offer the best combination of strength, light weight without warping and colour variation to be found in one stock.

McMillan offers a vast range of synthetic stocks, including my favourite Lazzeroni thumbhole designed modelled after the much-loved Harry Lawson Cochise.

We are currently spoilt for choice by the variety of synthetic stocks available, but most still come from abroad. The best-known names are such firms as McMillan, Robertson, Bell and Carlson, HS Precision, Christensen and Manners, but these have been joined by new firms like Staffordshire Synthetics and several English firms producing carbon fibre stocks that are making inroads into markets formerly dominated by American products.

McMillan

Gale McMillan, founder of McMillan Firearms, began producing fibreglass stocks in 1973. McMillan is now one of the most respected companies in the field, producing a comprehensive range of synthetic stocks. Tactical, hunting, benchrest, competition and ultra-light applications are catered for. Stocks can be made from fibreglass or graphite cloth, depending on the weight and style of the stock required for your custom project. They are usually made to order and can take six months to arrive, so this should be one of the first things to order for your custom rifle project.

Bell and Carlson

Bell and Carlson offers a good range of stocks at competitive prices. As well as building replacement or aftermarket stocks, they also manufacture thousands of stocks for original equipment manufacturers such as Weatherby, Remington, Winchester, Browning, CZ-USA and Cooper Arms. Bell and Carlson stocks are constructed using a hand lay-up process that uses composite materials of varying levels. These commonly include fibreglass, aramid fibres, graphite, epoxy gel coats, laminating resins and polyurethane reinforcement with milled fibreglass. This provides a warm and solid feel that is far better than you get from injection-moulded stocks. They are available in a good choice of styles and lengths as well as finishes. The integral bedding block/backbone stiffener option is a good choice for those concerned about extreme accuracy in all weather conditions.

The wide range of stocks for all types of rifles made by Bell and Carlson includes this tactical-styled model.

HS Precision

This is another American firm that has been producing composite stocks for more than thirty years. This was the first company to produce hand-laminated composite rifle stocks that also had an internal aluminium bedding block enhancement. The stock actually starts with these blocks, which are processed by a CNC machine from a solid aluminium billet and profiled to accept the action perfectly. The block is then placed in a mould of the correct style that has been lined on both sides with a hand-laminated layer of Kevlar, carbon fibre and fibreglass material. The mould is then injected with a polyurethane foam mixture that has a dense overall consistency. The stock is then finished externally and painted to the colour or texture of your choice.

HS Precision makes very good synthetic stocks with an internal aluminium backbone for rigidity.

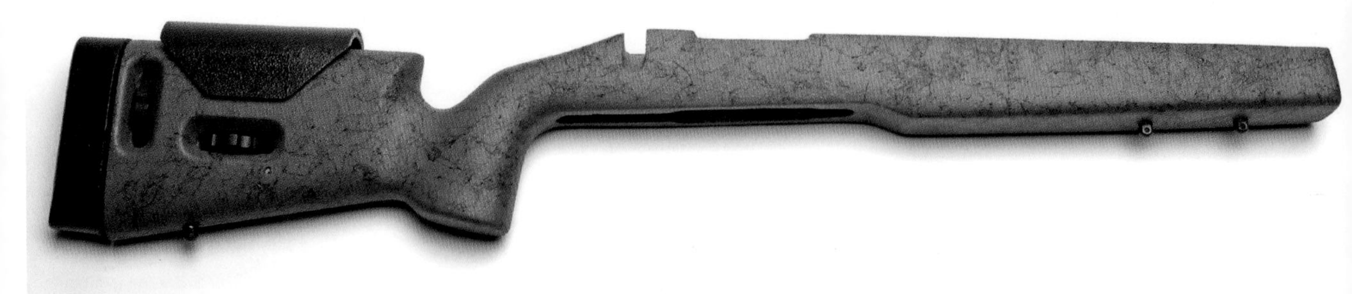

Carbon fibre stocks, such as this Christensen from Dolphin Gun Company, are becoming popular as they are light but strong.

Christensen Arms

Carbon fibre stocks by Christensen Arms, available in tactical adjustable, sporter and thumbhole styles, are less than half the weight of other tactical rifles and are very comfortable to shoot. They are hand laid and the natural carbon fibre, cross-hatch exterior finish is both attractive and functional.

Manners

Manners stocks are designed to be the most rugged, stable, accurate and ergonomic products available without being overly heavy or bulky. The outer shell casing is made with 35 per cent aircraft grade carbon fibre and 65 per cent fibreglass in multiple layers, hand laid with a high-temperature epoxy resin. This is performed in a vacuum and then cured using heat to maintain a uniform interwoven layer. A set of pillar beds are inset and the correct action machined into stock with a gel coat outer layer, which is available in a variety of colours and grip textures.

The Elite Hunter series was developed to be as light as possible while maintaining the ruggedness, stability, accuracy and ergonomics of a proper hunting stock. These stocks have a 100 per cent carbon fibre outer shell with a very light inner fill in the areas that do not need reinforcement. Then weight is added with a synthetic material and pillars are inserted with reinforcements around the recoil lug/front pillar and web area in front of the trigger. Models can be ordered with all bottom metal and a detachable magazine system that solves the problems that arise when buying options from differing manufacturers and finding they do not fit. Manners offers this as a standard option. An optional

Stocks made by the US company Manners are also light and very strong. If desired, they may be supplied as a complete package with bottom plate and magazine.

Carbon fibre can be painted, enabling the striking effects shown on this custom gun from the Anglo Custom Rifle Company.

folding sporting stock is also available that enhances its popularity on tactical rifles, since it allows the stock to be folded away by pushing a small button in the pistol grip area and swinging the butt section to the side on a precision-made hinge until it folds flat against the left side of the stock for easy storage in a travel trunk.

METAL STOCKS

Some stocks used for longer-range shooting use a modular metal or part-metal framed stock. This allows several parts to be swapped, replaced or fitted to differing customers' sizes, preferences or disciplines of shooting. Often when aluminium is used it makes for a lightweight stock, but areas that you hold or where your cheek touches the stock need some form of comfort, such as wood, plastic or soft-touch coating pieces.

Accuracy International

These stocks have an aluminium chassis onto which a polymer exoskeleton is bolted. This gives the advantage of a rigid bedding area, yet the external appearance can be moulded to your own liking. The AW AICS version is the

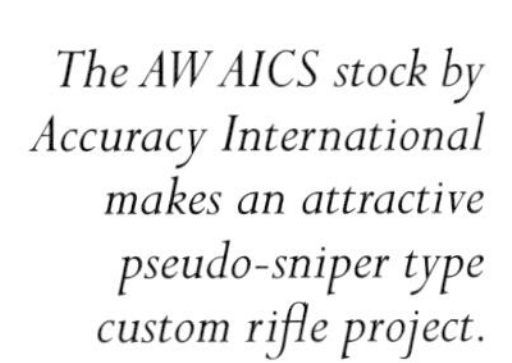

The AW AICS stock by Accuracy International makes an attractive pseudo-sniper type custom rifle project.

The AX AICS folding variant and Predator action on a Kershaw custom rifle makes a practical short rifle for most forms of shooting.

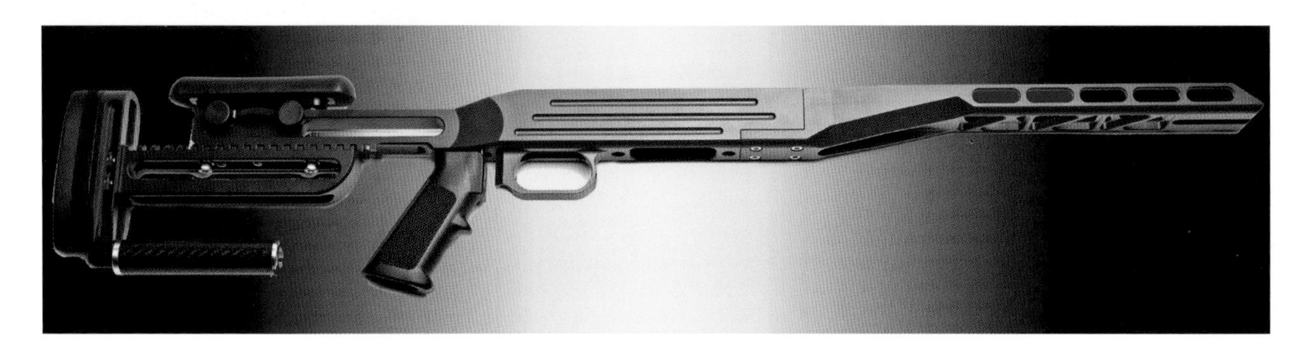

Dolphin Gun Company's in-house CNC machined, aluminium-framed chassis-type stock is superb for target shooting, long-range varminting or just hunting.

thumbhole variety seen on their sniper rifles. It is popular for tactical-styled custom rifles and is available in rigid or folding stock varieties.

The AX AICS version has a pistol grip arrangement with aluminium exoskeleton and vented forend, with Picatinny rail attachments for accessories such as lasers, night vision, bipods and lamps. This is also available as a folding stock variant.

Dolphin Gun Company

Dolphin makes its own aluminium-framed chassis, which can be inletted to a variety of actions. These stocks are all made in-house on one of the many CNC machines with two style stocks: one for F Class T/R and the other for tactical rifles. These stocks are modular in design with full inlets for Remington 700 style actions, Barnard S & SM Actions and RPA Quadlite, both single-shot and repeater, and the BAT VR action. Fitting for Savage actions is also to be introduced.

The stock is fully adjustable, including an adjustable pistol grip for length of pull, with a choice of three types of forend: short tactical style, long F/TR style and a Benchrest F Class Open forend. The repeating version comes with a five-round AI magazine as standard and is fitted with a fully adjustable Morgan recoil pad and a quality pistol grip. All stocks come complete with stainless steel bedding bolts. A special feature is that Dolphin can inlet any of these stocks to your rifle's action on their CNC machines. They are also available in a variety of anodized colour finishes or coated in-house with Duracoat in your choice of colour. These can be a straight inlet to the stock or, if preferred, a separate bedding block can be machined and inletted that perfectly mimics your rifle's action.

STOCK PROFILE AND FIT

Having chosen whether your stock material is to be wood, laminate or synthetic, your next decision is to choose the style or profile. The final use of the rifle will dictate the style in most cases, although custom stock makers also supply transitional stock designs that span several uses to make their products more versatile. Most custom stocks used on rifles in Britain are aftermarket synthetic or laminate stocks as they are cost-effective, hard-wearing, practical and easily bedded correctly for accuracy. For the majority this is fine, but then most custom rifles start to look the same. True custom stocks, however, are still produced from scratch in the traditional way. This especially applies to traditional classic sporting rifles, while others are crafted by hand from laminates in more modern designs. There are many variants on the basic design types, but these are the most common.

Sporter

This is the most common style. It utilizes a slim forend with trim contours and is designed to fit the great majority of average shooters. On a custom rifle you can keep the sporter ethos and specify differing forend styles, like the Schnabel, bird's beak, bull nose or slanted. The rear butt section can have an ambidextrous feel without a cheekpiece look or a one-side only cheekpiece with a European styling, old English dropped style or Monte Carlo raised type. You could even have a combination of these – that's the fun of a custom rifle. Thumbhole versions are my favourite.

Varmint

Most varmint rifles sport a heavier barrel for long range and prolonged use. The stock design reflects this: usually the forend is wider to accommodate the larger-profiled barrel and can have a beaver tail profile that allows a wide grip or the use of a bipod for a stable platform. The rear butt section can be any style you like. A bias to allow the bottom section to fit into a rear rest, however, helps to keep it stable when using shooting bags. Varmint rifles are available in any configuration of laminate, wood or synthetic, but they are usually biased to fit heavier barrels than their sporter weight cousins.

LEFT: *The Sako 17 PPC in typical sporter-type stock is both light and slim in profile.*

BELOW: *Sako custom .22-250 in varmint profiled walnut stock with a wider forend to fit a bipod and use a heavier barrel profile.*

The Classic design is elegant with flowing lines, a long raked pistol grip and shadow-lined cheekpiece.

Classic

Classic custom rifle designs have simple yet elegant lines. Uncluttered cheekpieces are usually dropped for a straight line comb and slender profile. The pistol grip, too, is finely shaped and has a longer rake than normal, while the slim forend usually has a traditional classic rounded nature. The cheekpiece traditionally has a shadow line under it to emphasize the shape and the flat-cut butt can be fitted with either a curved steel butt plate or a recoil pad. The pistol grip looks good with a steel grip cap either being full, faceted or skeletonized, as with the butt plate. The understated or simple, uncluttered, clean lines have a lasting appeal among custom gunsmiths and shooters alike. This style is also well suited for British express sighted rifles where shooters like to use open sights on their rifles.

Tactical

A tactical or semi-tactical stock will usually have some form of adjustment to the stock, either to the cheekpiece or the butt section. These types of stock are becoming more popular as long-range and CSR rifle shooting demand more custom rifles configured in this way. Where do you start? Usually with a synthetic stock but with the addition of a fully adjustable, often skeletonized butt section with a pistol grip or thumbhole. The action usually has a separate bedding block and the forend looks like a Meccano scaffold crane with more attachments for lights, lasers and night vision imagers than the Blackpool Tower.

Benchrest

As the name suggests these rifles are shot off the bench. The commonly used term 'riding the bags' refers to a stock with a flat wide forend and a parallel bottom section to the butt stock. When the rifle recoils it moves back unimpeded and consistently, which is the aim of any benchrest rifle. The stock type can be wood, synthetic or laminate, but synthetic is usually more popular. Skid plates and skid rails may be added, or stocks might be filled with lead to reduce recoil

Tactical-styled custom rifles, such as this example by Riflecraft, are popular for long-range shooting, but they can be used for any type of shooting.

Benchrest-styled stocks are normally synthetic or laminate. They have flat forends and rear butt sections that slide or ride the bags used as rests.

on extra-long range rifles in order to achieve the smallest possible groups downrange.

CHOOSING A PROFESSIONAL BEDDING SYSTEM

Why do you need a professional bedding system when you have a custom rifle built and why does it cost so much? The easy answer is that, no matter now stable the stock might be, whether it is wood, laminate or synthetic, a good cohesion between the metalwork (the action) and the stock is necessary to maintain extreme accuracy, precision and consistency shot after shot. On some traditional custom rifles bedding is still carried out in the traditional way with a painstakingly perfect metal-to-wood fit. This takes plenty of skill but does not always account for moisture changes to the stock in differing climatic situations that can shift a rifle's zero. Most custom rifles now use some form of syn-

Rob Libbiter is a master at crafting stocks from any type of wood, laminate or synthetic in any style of your choice.

Bedding is crucial not only for strength but for consistent accuracy. Here Derek Clifford inlets a stock for Steve Bowers.

thetic bedding compound to marry the action to the stock for a perfect fit.

There are many benefits to well-selected bedding:

- It strengthens the action area and allows heavy recoiling rifles to be shot with confidence.
- It avoids unnecessary stresses on firing caused by the rifle's action torquing in the stock under recoil, and thus avoids cracks, shakes and movement.
- It ensures that during and after the firing cycle the action and barrel returns exactly to the same position, thus increasing the shot's consistency.
- Zero is maintained under any adverse temperature, moisture or impact.

Ideally bedding takes several forms, with the most common being the removal of wood or synthetic material around the area that comes into contact with the action in the stock, This is replaced with an inert stable material like fibreglass or epoxy resin-based products such as Devcon, Marine-Tex or Acraglas. This then forms a skin between the metal and stock and makes an exact footprint in which the action sits.

Bedding can take many forms from simple glue-in jobs to pillar beds, where the action sits on stable aluminium pillars through the stock. Most wood, laminate or composite stocks can be bedded, but injection-moulded hollow stocks are rarely suitable. Some traditional makers still like to bed the entire barrel and action, which requires a constant pressure along its entire length. Unless this is done, heat build-up from firing multiple rounds can cause the metal and stock to shift slightly, changing accuracy and the rifle zero. Most shooters opt for the action-only bedding method in which the action is securely cradled in a synthetic bedding compound, with or without pillars. This allows the barrel to float freely so that on firing the harmonics are unchanged and consistent shots will be achieved. Separate bedding blocks that are matched to the action and then inserted and glued provide a stable base to ensure the action does not move. Some shooters, especially long-range aficionados, use bedding blocks that secure the barrel close to the action and allow the majority of the barrel to flex on firing along with the action. So long as they flex freely the same way this can offer good accuracy.

It is also important that bedding helps strengthen the stock as more and more custom rifles are being fitted with large, long and heavy profiled barrels. If these are only held down by the stock screws the leverage on such a long, heavy barrel soon causes a lot of pressure and stress at the action end, which is detrimental to accuracy. A good bedding job provides a solid base to support the whole rifle and bedding just forward of the recoil lug and the first two inches of the barrel can also be helpful in these cases.

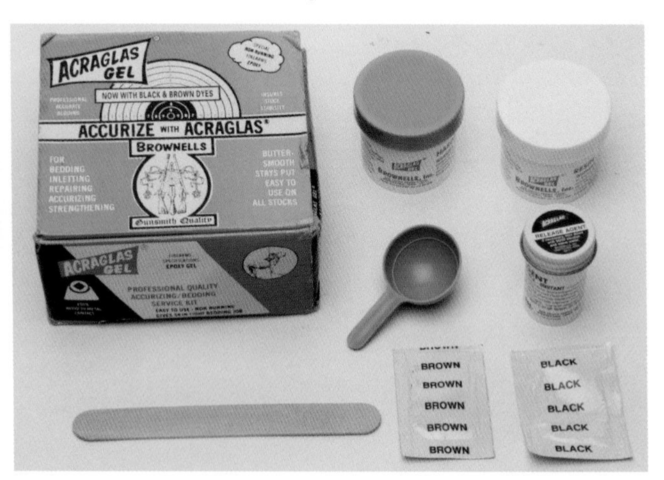

Acraglas is a good bedding compound, but Devcon and Marine-Tex are now more widely used.

It takes skill to remove excess wood around the action inlet and recoil lug area for a perfect fit.

Glass Bedding

This is simply the process of substituting a portion of the stock where the action sits with a slow-drying epoxy resin that forms a hard surface to which the action is bedded. The compound used varies but it must be fully cured or hardened, devoid of airspaces, it must not shrink or expand due to climate change and it has to be chemically inert so that cleaning fluids from the rifle do not soften it.

This is the method Steve Bowers uses on his custom rifles:

- Remove excess wood around action inlet and recoil lug area.
- Fill the magazine well and the stock screw holes with plasticine to stop the bedding compound from entering these areas.
- Tape the sides of the stock to stop overspill attaching itself to the wood or synthetic surface.
- Mix the bedding compound with the correct quantities of filler and hardener.
- Coat all the metalwork of the gun that will come into contact with the bedding compound with release agent to prevent becoming glued in permanently.
- Introduce the barrelled action to the stock,

Plasticine is used to fill the magazine well and stock screw holes to stop the bedding compound hardening in them when the action is attached.

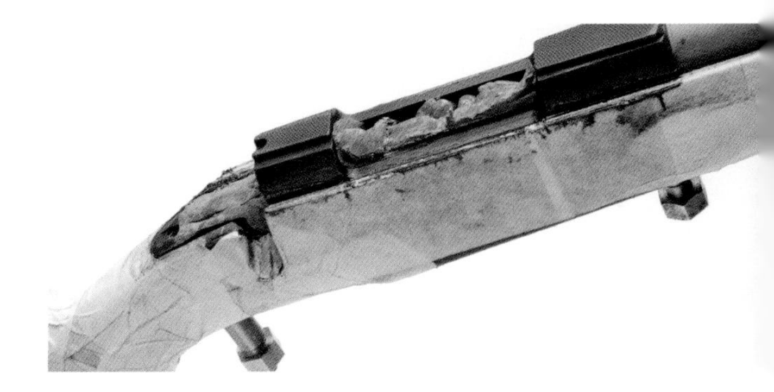

Unless you tape off the stock, the bedding compound can be hard to remove.

Specially custom-made stock screw tighteners ensure the action has the correct and even tension when bedded to the stock.

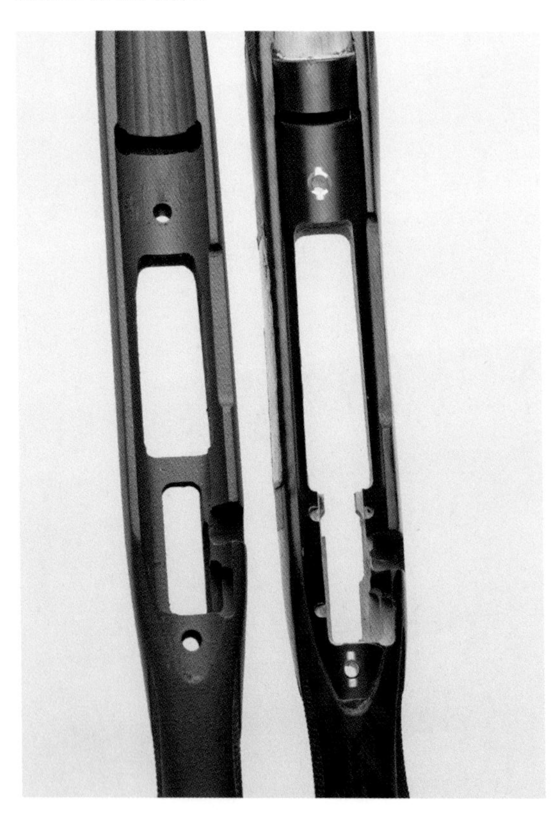

Bedding a sporter stock will ultimately result in a more accurate custom rifle.

secure it to the normal depth and allow the compound to harden.

- Special stock screw tighteners can be fashioned as desired.
- Remove and check the surface for uniformity. If it is not flawless, you can skim and rebed just the top area to a mirror finish that is an exact copy of the metalwork.

Pillar Bedding

This is the icing on the cake and should be used in conjunction with the bedding compound. The pillars are two metal tubes drilled to be inset where the stock screws go. They are oversized and sit flush to the bottom metal, that is to the magazine and to where the action sits. The stock screws then go up inside the pillars and tighten the action down onto the stock, so avoiding any crushing at this pressure point. They allow a very precise fit and uniform tension to be maintained between the action and stock. This ultimately translates to better accuracy and consistency as the action is rock solid and cannot move. It also means the barrelled action can be taken out of the stock and returned with little alteration to your zero, which is handy when you need to fit your rifle into a small travelling case.

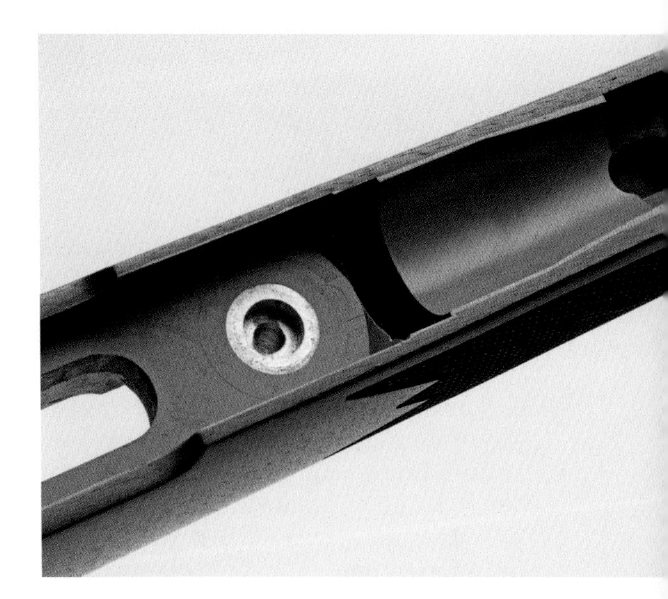

Pillar bedding uses aluminium pillars to prevent the action screws crushing the bedding compound.

Originally made of steel, most pillars today are aluminium to save weight but still retain structural integrity. They have the advantage that they can be used in wood, laminate, plastic or synthetic stocks, and their vibration-absorbing attributes are beneficial to all types of stock material. When used in conjunction with a bedding material you have the best type of bedding system for your custom rifle, but it must be fitted by an expert gunsmith to get the pillars parallel, otherwise all your good work is undone as the stock screws will pull unevenly, causing non-harmonious vibrations on firing, destroying accuracy and potentially damaging the stock.

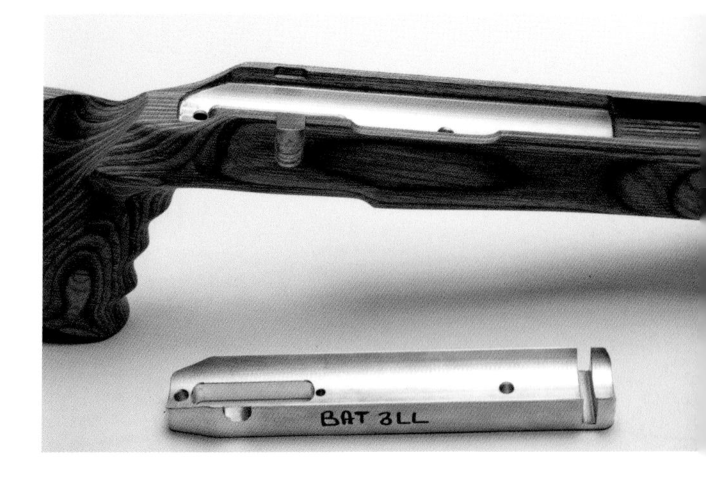

Dolphin Gun Company makes its own aluminium bedding block, which replaces the inlet of the stock and mirrors the underside of the action. This strengthens and increases potential accuracy.

Block Bedding

Another way gunsmiths overcome the bedding issue is to use a bedding block as an integral unit, replacing the inlet of the stock with an aluminium block that mirrors the underside of the action. When inserted into the stock the action sits on a perfect likeness of itself and a uniform bedding is achieved. Additional pillars can then be inserted to add stiffness to the action area. Aluminium is usually the metal of choice as it is light, easily machined, does not rust and is easily milled to get a perfect fit. I have seen titanium blocks, but these very expensive, while stainless steel is too heavy. Delrin, a homogenous synthetic material, is sometimes used.

Action and Barrel Bedding Block

Taking the bedding theory a little further, many shooters believe that the best system is a bedding block that fits in front of the action, gripping the barrel at its fattest profile, and is then bedded to the stock. Here a custom-machined billet of aluminium is sectioned and perfectly radiused to match the contours of the barrel profile. It is then securely clamped down and

Steve Bowers makes superb bedding block rifles that clamp around the barrel and can float the action or be used with the action to increase the bedding area on longer-barrelled rifles.

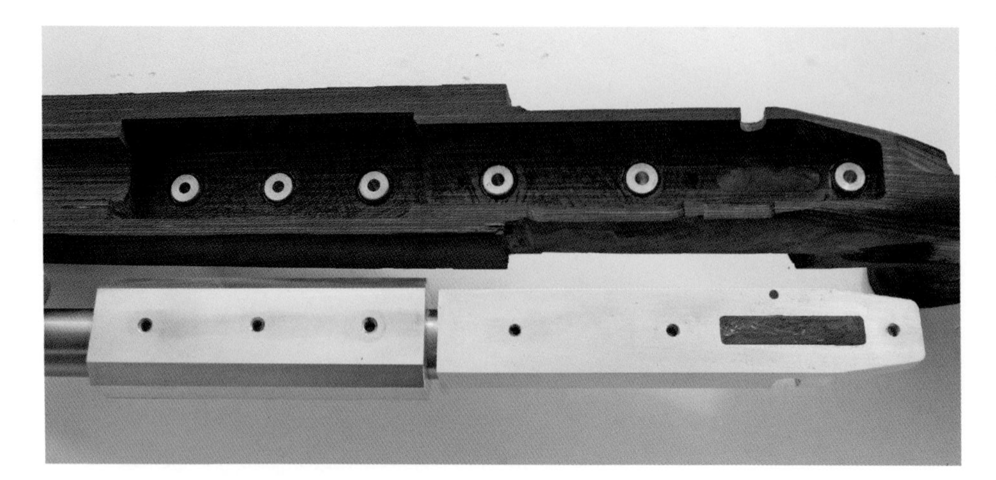

A Steve Bowers six point bedding block rifle, ready to win another long-range shooting competition.

fixed via pillar bedding through the stock. Steve Bowers uses this system on top-of-the-range, extreme-range precision rifles for his discerning customers.

The theory is that the enormous bedding block controls the pressures at the most hectic part of the rifle, the chamber where all the combustion processes take place, leaving the barrel and action to vibrate freely in a natural state without any pressure on them. This is thought to allow a perfect and, more importantly, consistent resonance to occur, greatly improving accuracy. The drawback is that this system is best used on large, long-range rifles that can accommodate such a large bedding block and a suitably sized stock.

STOCK ENHANCEMENTS

The beauty of an oiled stock cannot be underestimated. This is often is the only embellishment that it needs.

Traditional Oil Finish

If you want a tough yet sympathetic finish that brings out the figuring of the wood, but is resilient enough to shirk off a few knocks and the inevitable British weather, an oil finish is very good. Oil finishing kits such as that available from CCL allow you to prepare and finish a stock to a semi-professional level. Do not try to rush it. It is indeed very rewarding to see the fine walnut stock come to life and doing it yourself enhances your special feelings of ownership of your rifle. Every custom gunsmith has their preferred method of oil finishing, but here is one way that works for me.

CCL offers a kit that has three main components: red root oil (oil of alkanet) to bring out the figure and colour of the wood; the oil hardener to fill and give a natural looking finish; and the polishing compound to prepare stock for the oil and polishing the surface between applications.

- Prepare the stock by wetting with water and then dry to raise the grain. This can then be sanded down with successively finer abrasive paper. This will give a smooth finished stock surface ready to receive the oil, but if the grain is open a sealer may be necessary to fill the wood pores and smooth down again with the grading grit.
- Apply the alkanet oil to enhance the grain and colour and allow to soak overnight. Keep applying until the pores are sealed and the stock takes no more alkanet.
- Apply some of the hardener and apply every day to harden the oil already soaked into the stock.
- Rub down the stock with polishing compound to remove built-up high spots of oil.
- Apply equal quantities of alkanet and hardener to the wood until it starts to harden and feels tacky.

RIGHT: *A traditional oiled finish can be achieved by using the CCL stock finishing kit, as here on my Sako Finnfire custom stock.*

BELOW: *Lacquered stock finishes are very practical to seal and use. They are particularly suitable on laminate woods, for example on this Tikka 6mm BR custom.*

- Remove excess. Using your palm, rub the stock so that the heat of the hand cures the oil/hardener mixture.
- Repeat the previous stage until the stock takes no more oil/hardener mixture.
- Finish using the polishing compound until the surface has a high lustre.

Lacquer

Lacquer is best applied by a spray gun for an even layer that can be left as is or rubbed down and reapplied for a deeper lustre. Whether gloss, matt or satin finished, a lacquered stock can be applied to all stock types. You should be aware that on a walnut stock it can look a bit artificial and when scratched shows up a lighter colour that mars the overall finish and is not easy to touch up. It actually enhances the finish on a laminate stock, however, bringing out the colours of the laminate beneath. It has the advantage of being cheap to apply, but the additional weather protection and non-moisture ingress is very beneficial for rough use. Paint, custom wraps, Cerakote and Hydrographics finishes, which cover both stock and metalwork, will be discussed in Chapter 7 below. These types of finishes are becoming increasingly popular and shooters are using ever more elaborate stocks of differing or mixed materials.

Chapter 2
Rifle Actions

Just as you wouldn't build a house on poor foundations, so you should not build a custom rifle on a poor action. The action is the foundation to which all the other principal components of your custom rifle are attached: the barrel, stock, scope, trigger and magazine arrangement. Get the action wrong and all your hard work sourcing the best components will be wasted. The action is pivotal to any custom rifle project. A custom rifle built on a factory action with a new match grade barrel and no other alterations will shoot very well. The next step up is to have a factory action 'trued' to cure any imperfections left at the factory manufacturing stage. A full custom action, built from scratch to exacting standards and tight tolerances, however, will give the best possible stability and function, enabling a custom rifle's other components to play their part.

The most common action, and probably the most practical, would certainly be a repeating bolt action design, but many good custom rifles are built on a single-shot, lever or straight-pull action. The choice of type is really down to the client. Both falling block and break single-shot actions can offer something different and have an old-school air of sophistication. Lever-action and straight-pull designs are less common, but I have seen good lever-action customs made for close-range boar shooting where a fast second shot is necessary. Similarly the straight-pull bolt action is very fast and rifles like the Lynx 94, produced in small numbers in a Finnish factory to clients' individual requirements, are almost custom from the off.

However, it is not as simple as that. The custom rifle industry is flourishing, spurred on by a rise in the popularity of deer stalking and long-

Custom rifles can be made either on a factory action or from a purely custom-made action, depending on your budget.

Variants of the Mauser 98 bolt action remain the basis of most custom rifles today.

range shooting, each discipline creating a need for better competition rifles. There is seemingly an unlimited combination of specifications to sieve through in order to achieve the custom rifle of your dreams. In some ways that is good as you can get exactly what you want, but it can be a minefield to the uninitiated.

ACTION TYPES

Factory Actions, Repeater

Mausers

The Mauser 98 action has spawned more of today's modern designs than any other. One of the enduring features of the Mauser design is that it was initially meant for military use, which required a strong lock up, controlled round feed, reliable extraction and ejection.

The non-rotating, one-piece extractor was a big factor in the Mauser action's popularity for sporting use as it ensured the cartridge was 'controlled' all the way from magazine to chamber and then out again to be ejected. This was very handy on early rifles destined for hunters in Africa up against a charging Cape buffalo. This design has been copied on the Springfield '03, Enfield 1917 and Model 14, as well as the pre-'64 Winchester and later Ruger M77 rifles. Another interesting feature is that the extractor or hook section is held tight against the cartridge case at all times. This means that when a hard extraction happens all the surface area of the extractor claw is engaged on the case and does not pop off the rim, as with some rotating extractors. Because the extractor claw sits forward of the bolt face, more of the cartridge can sit into the rifle's chamber for safety and support. Less than a twelfth of an inch (2mm) of the case on a Mauser sticks out of the barrel's chamber, whereas on some rifles they protrude three times as far. Mauser actions and bolts are drop forgings, ensuring a uniformity in grain structure. They are made from a low carbon or

Mauser actions have always been used for classic custom rifles, such as this Jefferies.

mild steel that gives high quality and low impurity. These are then machined to shape. Critical areas such as bolt raceways and locking lugs are treated with carburizing paste and placed in a heat treatment oven. When cured these parts are then tempered to the correct hardness and form a 'case' around the action with the milder or 'softer' steel beneath.

Most modern Mauser type actions are made from chrome molybdenum alloys and are thus hard throughout, rather than the hard case surface of older Mausers. Early Mauser actions were made from carbon steel and then heat treated to harden the outer shell. Towards the end of the Second World War the outer thickness had been reduced to as little as 0.004 thou to 0.010 thou, but a figure of 0.015 thou should be considered the minimum acceptable. The bolt and action should test between 42–45 and 38–42 on the Rockwell hardness scale (RC) to avoid problems. It should be noted that Mauser actions are hardened where they need to be and not all over the action, which can lead to confusion when modifying an old Mauser action for custom work.

Older Mausers often have their bolts replaced, reshaped, safeties changed, clip magazine lips rounded off and new hinged floor plates fitted to start the custom rifle project. The standard length can be worked on and the bolt opened up to accept cartridges up to .375 H&H or 404 Jefferies, but true magnum length cartridges like the 416 Rigby or 505 Gibbs need a magnum length commercial action or special bolt head and magazine rework. Mausers are good to work on because from a custom point of view they offer endless opportunities to customize to your tastes. As you would expect, there is a huge variety of aftermarket parts available to enhance an old military action, including teardrop bolt handles, hinged floor plates, safeties, single-stage triggers and scope mounts. A custom gunsmith would be needed to install more advanced accessories, such as adding square bridges for scope mounting to the tops of the action, as on the original Oberndorf Mauser sporters. These look superb and allow the use of claw-type mounts for a quick detachable scope, which are popular in Europe and on African

David Lloyd rifles were way ahead of their time. They incorporated a Mauser action and were fitted with his own special scope mounting system for no loss of zero.

rifles. Argentine 1909 actions made by DWM are highly regarded, as are the smaller receiver ring G33/40 Mausers, which make a great lightweight deer gun. When considering an old Mauser 98 action you need to decide whether it is worth spending thousands of pounds on it when you could buy a new custom-made Mayfair action for about the same. Many good British custom rifles have been made on a Mauser action, such as those by Rigby, Westley Richard, Holland and Holland, Jefferies and the small independent customizers, among them Ron Wharton, TT Proctor and Medwell & Perrett. There is also the David Lloyd rifle, an iconic custom rifle from the 1970s with 'Back to the Future' looks that made it the DeLorean of its day.

If instead you choose to go for a modern Mauser design, most of the hard work has been done for you and it can be customized to your heart's content. Old war actions can be both good and bad propositions for conversions and more modern versions may offer a better option. Actions by Zastava, Parker Hale, FN or Interarms Mark X are all well made and good value. CZ or Brno offer large action sizes that are very popular with customizers for a classic large-calibre African rifle.

Remington

The availability of aftermarket custom parts has made Remington the most popular and copied factory, with Savage actions coming a close second. The Remington Model 700 is probably the most used or cloned action after the Mauser 98 type. Since its introduction in 1962, this strong sporter type repeating action has been available to handle calibres as small as .17 Fireball up to 416 Remington. It has drawn a large following and created a demand for aftermarket products that have further added to its versatility. Its rounded profile makes it an ideal candidate to 'blueprint or true' as it can be set up in a lathe very easily.

The bolt head encases the cartridge head. The extractor is the weak point in the design and is often replaced and remachined for a much more solid Sako type. The twin locking lugs are easily lapped to fit precisely and

Many custom rifle makers rely upon the Remington Model 700 action. It is strong and there is an almost inexhaustible supply of aftermarket parts available for it.

there is a strong camming action as the bolt is opened. It also has a fast lock time of some 2.9 milliseconds, reducing the point between pulling the trigger and ignition. The 40X version is a single-shot version of the Remington 700 with a far stronger rounded profile and larger surface area to bed into the stock. Without a magazine cut-out the action has greater rigidity and strength of 2.3×10^6 ($in^2 - lb$), compared to the 0.5×10^6 of a standard Remington 700. The action face is also squared and lugs lapped as part of the manufacture.

Sako

Sako is one of my favourites as the older model actions were made when only metal parts where used, all machined with precision and largely by hand without the use of the generic plastic used today. Sako was founded in 1919 as part of the Finnish Civil Guard as a minor rifle repair workshop with only eight employees in an old brewery in Helsinki. In 1927 operations moved to a bankrupt ammunitions works, with a 300-yard rifle range, in Riihimaki and became known officially as Suojeluskuntain Ase- ja Konepaj Osakeyhtio (Sako). During the 1920s and into the war years the models 28, 30 and M39 rifle were produced but it was not until two years after the war that a true lightweight hunting arm called the L46 was available. Extremely popular, it formed the basis of the design that Sako used over the decades, added to by a medium-action version to facilitate heavier calibres and then later, as the demand for big game rifles increased, a long-action version was manufactured.

The three sizes of action sizes are essentially a highly refined Mauser-type design made to Sako's impeccable standards. The receiver is a forging of toughened moly-chrome steel that is stronger than comparable investment casting methods. The bolt is another forging that has a one-piece bolt handle and bolt body design for strength with twin locking lugs up front that

The older A or L series actions made in Finland by Sako were some of the best factory actions ever made. Here an L691 action is used for a Steve Bowers 6mm AK custom rifle.

Sako painstakingly ensure make total contact with the bearing surface, with a small third lug at the rear on the Finnbear model as a safety lug. Ejection is via a mechanical spur that positively flings the case clear, aided by the often copied extractor claw. The integral recoil lug at the front of the action provides a good bedding surface for fitting it securely to the stock. The well-made, all-metal magazine is the fixed type with hinged floor plate, providing a reliable means of feeding cartridges. The receiver top has the Sako tapered dovetail integral scope rails, which provide low positive and recoil-proof scope mounting. Finally the trigger is very crisp, precise and adjustable, just what a sporting arm to secure every last bit of accuracy. The overall finish of all Sako actions is a highly polished blue with very few tooling marks, which makes them popular choices for any high-quality custom rifle project. I have used a Sako M 591 action for my .20 BR rifle from Venom Arms. The Sako fully silenced .308 Win with a Harry Lawson stock is built on an A2 action. A Sako 78 forms the basis of the .17 Ackley Hornet.

Tikka

Tikka is another excellent candidate for custom projects. The newer T3 models are good but again I prefer the older, all-metal models, as with Sako. Tikka has had a factory in Finland since 1893 but it was not until 1967–9 that the first true bolt action sporting rifle was manufactured, the Tikka M-76 (later named the LSA 55). This was a very popular, reliable and accurate firearm, but its action size could only handle cartridges up to the size of the .308 Winchester cartridge. Therefore a larger-actioned rifle was introduced in 1970 called the LSA 65 (later the M65), which, despite its similar name and use of the common stock and barrel, had a completely different action design. Again this was hugely popular and in 1972 a Sporter version was made for Biathlon events. In 1983, however, Tikka was merged with Sako Ltd and the Tikka factory was replaced by Sako in 1989 the Tikka factory name. Shortly after the M55 and M65 models were replaced by the newer M558 and M658 models.

The actions of the M65 and the smaller M55 were completely different, This was chiefly because the lower section of the M65 action body was fashioned in a round profile from a single billet of steel and had an integral recoil lug dovetailed into the underside of the action front. The M55 had a separate recoil lug, but the integral M65 lug was much better. The bolt too had undergone a makeover. Apart from being longer, it still retained the twin forward locking lugs but had a further two smaller lugs at the bolt's rear that acted as a safety feature when the bolt was closed. There are two exhaust holes along the bolts body to vent gases if a primer blows. The excellent Tikka integral scope dovetails in the same way on both actions. The long action configuration of the M65 rifle was ideally suited for the larger .30-06, .270 or 6.5 × 55mm cartridges as well as the 7mm Remington and .300 Winchester magnums. The M55 can be used for .222 to .308 Win or custom calibres like .20 Tactical, 22 Satan or 30-47 Lapua. The detachable magazine was also modified on the M65 model, which has a staggered magazine design that can be loaded from the top of the action *in situ*, making it easier to load in the field. Many old M55 and M65 rifles find their way into quality gunsmiths' workshops to be converted into very fine, superbly accurate custom rifles. I love them and have a Venom Arms .220 Swift Tikka M65 custom rifle, .338 BR M65 super/subsonic custom, M55 retro 7.92 × 33mm Kurz stalking rifle, 6mm Dasher long-range varmint custom on an M55, 7 × 57mm AK Improved M65 Callum Ferguson and, finally, one of thirteen custom M55s commissioned by GMK Ltd.

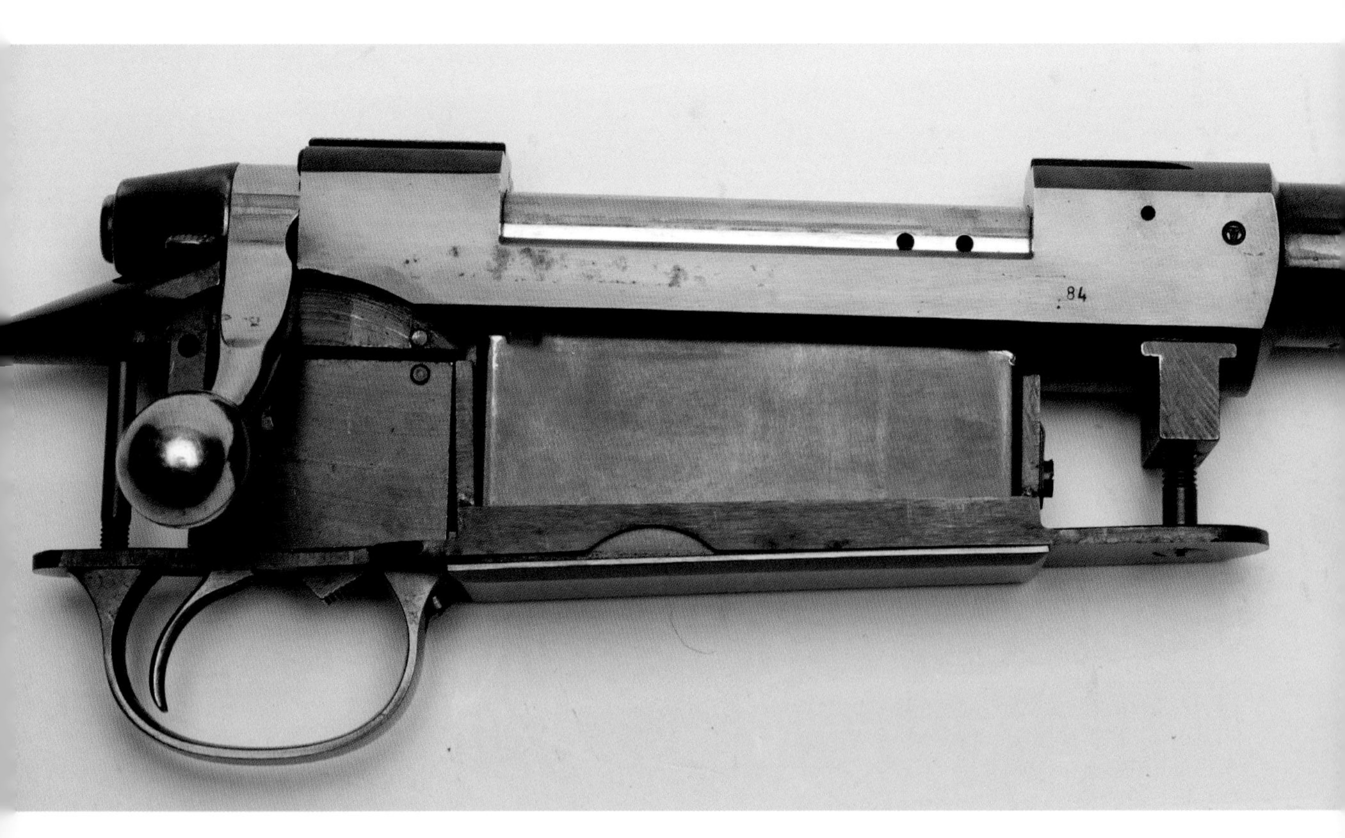

Like those made by Sako, Tikka rifles offer excellent donor actions for custom rifle projects.

RPA Quadlite

This is one of my favourite actions for testing new wildcats. You can instantly see the British-made RPA Quadlite's target pedigree in the action design. It is notably substantial and is available in left- or right-hand configuration. The main body is a full 6.5in long and made from high tensile steel with a super-hard satin black finish that offers good corrosion resistance in use. The rounded exterior profile is only broken by a slim and small ejector port. The front of the action face has a slot into which the separate recoil lug locates to resist movement; this is also substantial enough and square to aid in a solid bedding area. When refitting differing barrels these can be exchanged with other sizes to alter the headspace.

The bolt is a work of art, again made in one piece, 6.75in long and precisely machined. As the name Quadlite suggests, the bolt has four locking lugs to the front, arranged at 12, 3, 6 and 9 o'clock, with a plunger-type ejector sited in the 3 o'clock lug. This is passive in operation, which means it is non-sprung, but as the bolt is withdrawn a spigot in the action locates behind the plunger and ejects the case with a force equal to that applied to moving the bolt at the time. A large claw extractor is located in the 9 o'clock lug and grips the rim of the case positively. I use two RPA actions for most of my wildcat cartridge work due to their strength and their ability to be used as switch barrel rifles: it takes less than five minutes to change out a barrel using a Farrell barrel vice on the bench and swap it over for another. In this way the only expense is the cost of a new barrel, not a complete rifle. I use two differing bolt sizes, 0.378 small for .223 sized cases and 0.473 for medium 0.308 size cartridges, and a selection of stock configurations for differing hunting or varminting applications.

RPA, based in England, offers superbly made actions with a four locking lug system primarily for target use, as on this Quadlite action.

Factory Actions, Single-shot Custom

Ruger No. 1

The Ruger No. 1 rifle is one of my favourite-looking rifles and really lends itself to custom rifle use as a donor action. The Ruger No. 1 design is based on the old British Farquharson action, which utilizes a falling block action that gives a really strong lock up and is capable of handling calibres right up to 585 Nyati. The compact receiver means that the barrel length can be quite long without increasing the overall length of the rifle, making it a gunsmith's delight for custom wildcat cartridges.

The Farquharson-style lever locks directly into the trigger guard: a downward movement on this lever allows the breechblock to fall, giving access to the rifle's chamber. A final movement activates the extraction of a spent case and the extractor kicks the case out. As the cocking lever is raised the hammer system is fully engaged by the sear, the extractor cams back into position and pushes the breechblock back into position. A special feature of this system is that the breechblock itself moves slightly rearward as it is lowered, thereby eliminating drag on the rear of the cartridge and, conversely as the breech is raised, this movement aids in chambering difficult rounds. Because of the action's design, the No. 1 has a very good and strong extraction system that can cope with modern high-intensity rounds. The Ruger No. 1 is well suited to calibres with a nostalgic appeal, such as the .240 Apex or 7 × 57mm.

The side walls of the action lend themselves to engraving, gold inlay or colour case hardening, as does the reprofiling of the lever safety catch. The addition of custom quarter ribs and Holland and Holland sights make a superb custom rifle. Stock triggers are fair, but they can be lightened or replaced for double set models, if you prefer. The woodwork is where the Ruger No. 1 really shines. Its classic English shadow-line, drop cheekpiece design really suits this rifle with clean lines and great handling qualities. If you have a Ruger No. 1 custom rifle it is certainly worth having a few modifications made to the bedding system, since the upward pressure of the forend can be problematic as it sits on a forward hanger. Uneven pressure on the barrel is not conducive to good accuracy, but this can be rectified in several ways. I have a .14 Walker Hornet build on a Ruger No. 3, the No. 1's smaller brother with the same action but different cocking lever. The woodwork is from a No. 1. The forend has a tensioning screw that is tightened with an Allen key to adjust the upward pressure on the hanger and barrel. This is an effective way of fine tuning the accuracy. The woodwork is also shaved away where it touches the action face to avoid compression harmonics.

Ruger No1 or No3 actions, as on this .14 Walker Hornet custom rifle, are good choices for custom projects using a cartridge with a rimmed case design.

The quick-change stock and barrel system of Thompson Center Contender rifle actions are well suited to custom rifle projects, as on this Ivan Hancock custom special.

Thompson Center Contender

A quality single-shot rifle or carbine can be very appealing for a seasoned custom rifle shooter. The Contender is especially good as this rifle is capable of excellent accuracy and reliability. It is also possible to change calibres from rimfire to centrefire in a trice. Warren Center teamed up with the toolmaker K. W. Thompson in 1965 to produce a pure hunting arm that had its origins in a single-shot pistol for silhouette shooting derived from the Mexican Senderos in the 1960s.

Its single-shot design and the external hammer may be old-fashioned, and I suppose the break barrel operating action is a shotgun-like, but these features make it a challenge for a custom gunsmith. The action is incredibly slender for a rifle yet strong. This is primarily due to its positive lock up with the heavily sprung locking lug situated on the underside of the barrel. The barrel pivots on a hinge pin that traverses both the action and barrel location lug, tilting the barrel open to allow a cartridge to be inserted. To achieve this the curved trigger guard spur needs to be tugged rearward quite forcibly. One of the best features, surprisingly, is the external hammer system. These usually have a slow lock time or hammer fall, but this drawback is more than made up for by a switch on the top of the hammer that rotates 180 degrees. This selects between rimfire cartridges or centrefire use by repositioning two differing firing pins, hidden in the hammer's stem, that hit two inertia firing pins set in the action's frame. This simple feature allows Contender users to utilize an incredibly wide-ranging cartridge selection from .22 LR all the way up to the .45-70 and even .410 shotgun. The old models had pretty basic triggers that really needed attention by a gunsmith to get the best from them. At first only a very basic walnut stock without any frills was available, but their popularity led to a whole industry providing replacements in exotic woods, laminates or synthetics, and in all styles and configurations. My favourite, a custom item fashioned by Ivan Hancock from Venom Arms, wears exquisite Exhibition grade walnut and fine line chequering, colour case hardened action and custom 7-30 Waters barrel. It is a testament to the Contender's design that many owners are willing to spend a lot of money on all manner of custom items. A newer version is known as the G2, and the Encore accepts larger high-pressure rounds.

Lever Actions and Straight Pulls

While it has never really been iconic in Britain, unlike its image in the American heritage, a custom lever gun used as a short-range wood rifle has a special place, with Winchester and Marlin being good sources. A custom Marlin 1895, owned by a friend and remodelled by Callum Ferguson in .45-70, is a hell of a wild boar and close-range game gun. The lever action is a strange sort of mechanism and the tubular magazine is sited underneath the barrel itself. Both these areas need attention from a gunsmith to provide a smoothly reliable and functional rifle.

The external hammer system again is characteristic. As the lever is pushed forward the bolt body is cammed backward to cock the action and thus the hammer. It is a bit old fashioned and leads to a long lock time. Again Callum lightened the springs and honed the sears. The lock time is now faster, smoother and the accuracy has improved threefold. The trigger is similarly modified for a crisp and light break and the muzzle has been threaded for use with a sound moderator for subsonic bullet use. This makes for a quirky deer rifle, but it still has the ability to shoot very good groups on paper and harvest venison from the woods. It could also stop a charging Cape buffalo.

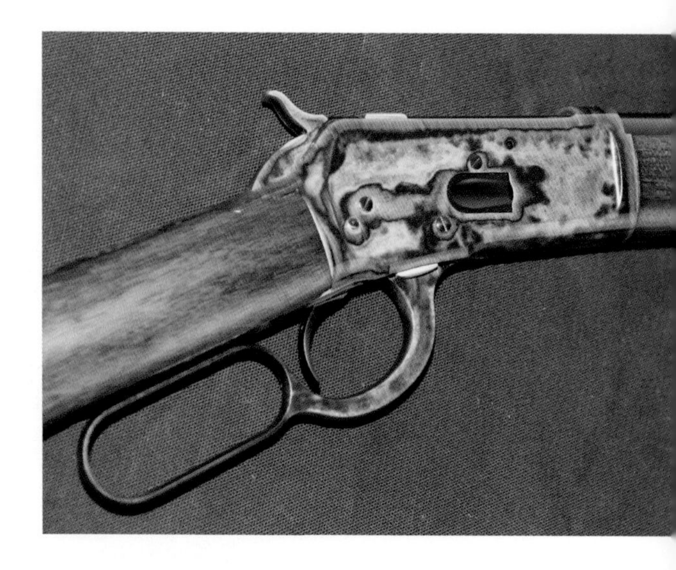

Norman Clark's superb colour case hardened Winchester.

A lever-action .45-70 Callum Ferguson tuned special.

Any discussion of straight-pull actions is really about Blaser, Titan 16, Heym, Merkel, Strasser or Lynx. These are factory-built rifles but the actions can be used for custom work, although the design means that many might be considered as semi-custom anyway. This is certainly the case with the Lynx 94, which is hand built in small numbers in Finland. Its straight-pull action would make a lovely fast-handling game rifle.

TRUED ACTIONS

Truing an action is basically a process where you take a factory action and make all the surfaces and dimensions to a concentric and correct alignment to one point, as well as making sure every part is square and true. Most factory actions are mass produced and are made to a certain tolerance. This means that barrel threads can

Trued Remington actions can form the basis of a good entry-level custom rifle, but to true an action properly, as on this Steve Bowers 20 Tactical, takes a lot of work.

be un-square to the action, with the barrel tenon threads not seated perfectly with the action's threads. Another very important area where the barrel needs to seat perfectly square is where the barrel contacts the action, otherwise it will exert an irregular pressure that on firing can upset the barrel harmonics and thus destroy accuracy and consistency. Where a recoil lug is inserted between barrel and action, again these items can be machined out of true and will certainly benefit from being honed to a flat surface to both sides. If it forms part of the bedding of the action, the lug also needs to be a uniform dimension to fit accurately into the bedding compound of the stock. The action itself would need to have threads in the receiver ring recut and trued to the outside circumference of the action body so that they are concentric to each other. This again will ensure perfect alignment. Many custom gunsmiths will spin up an action on a lathe, if circular in nature, to remove excessive metal and true up the action. If the action is square the same process can be obtained with a milling machine. In either case you should aim for as close a tolerance as possible.

The bolt that rides within the action will also need a lot of work on a factory rifle. The worst area is at the head end. Here the face of the bolt benefits from being perfectly square. If any modification is to be made to the bolt head, for example to fit a wildcat cartridge of differing head dimensions, this is the time to do it. Some bolts have poor extractor claws and many custom gunsmiths will remove the factory extractor and retrofit a Sako or M16 type large claw that sits within one of the locking lugs. These locking lugs, too, are of great importance as if they are unevenly machined at the factory the bearing surfaces on the back edge will not contact in unison, and thus the bolt will tend to lock up out of square in the action. This will put an uneven pressure on the cartridge case and place the alignment out of true.

While truing the bolt lugs it can be very beneficial to install a lug radius bearing to support the outside edges of the bolts, ensuring that the lugs sit perpendicular to the barrel and they lock perfectly flat in their recesses of the action.

Typical Action Truing Procedure

The desired effect of this truing or blueprinting an action is to remachine the action so that the bolt raceway, locking lugs, receiver face and threads are all perfectly concentric and square with each other and to the barrel. You can also sleeve the bolt, which removes any radial movement or play in the bolt when closed. The sole purpose of remachining is to improve accuracy, since eliminating as many defects and anomalies as possible will bring you as close as you can get to a perfect action. Realistically you are aiming for a solid and concentric barrel to action union, which stops any flexing between the joints or threads *in situ* and, more importantly, under recoil. If the bolt wobbles then those vibrations will be transposed along the action and down the barrel under firing, causing 'bad' harmonics that interrupt the normal vibrations for a fired bullet and resulting in erratic shots and poor accuracy.

Every custom gunsmith has their own way of blueprinting an action. The following description of one such method uses a standard Remington Model 700 action by Steve Bowers.

The first step is to establish the correct reference point from which the other truing procedures are taken. Start by truing up the raceways in the action to make them concentric and parallel using diamond lapping paste applied by hand. It is important here not to twist or clamp an action straight from the outside, otherwise the inner raceways will only distort back when the pressure is released.

Next fit an arbor in place of the bolt so the action can be centred in the lathe. Make sure the arbor is a tight fit and protrudes from both ends in order that you can clock up and centre accurately with a micrometer gauge.

Machine a datum face on the action. This is simple on the flat bottom of a squared action, but on a round action, such as a Remington M700, you will need to lathe or grind the total outside diameter concentric to the raceways.

Put the action on the lathe using a fixture (if non-rounded) or collet (if round, as the Remington 700). Dial up the gauge to check it is running true to within 0.0005in.

Remove the arbor and true up the abutments or bolt lug seats by removing only enough metal to seat the lugs correctly.

The threads where the barrel is secured to the action are now recut, as factory threads are usually sloppy and not straight. This is extremely important for the highest accuracy.

It is essential that the action face is perfectly square with the newly trued threads from the action to the barrel. Machining this square ensures the barrel sits totally flat to the action with even tension.

The recoil lug is now checked for parallel faces and ground perfectly flat, ensuring that no tolerance remains. Unless this is done the action will not be bedded well in the stock and a perfect union will not be achieved.

Truing up the raceways. Here Steve Bowers has inserted a new bolt lug collar.

Re-cutting the factory action threads to match the barrel tenon threads, so ensuring perfect unison.

Now turn your attention to the bolt itself. Having trued the raceways, it is essential to make the bolt a tight fit within these to stop any movement that would destroy accuracy potential on firing. Custom gunsmiths often sleeve a

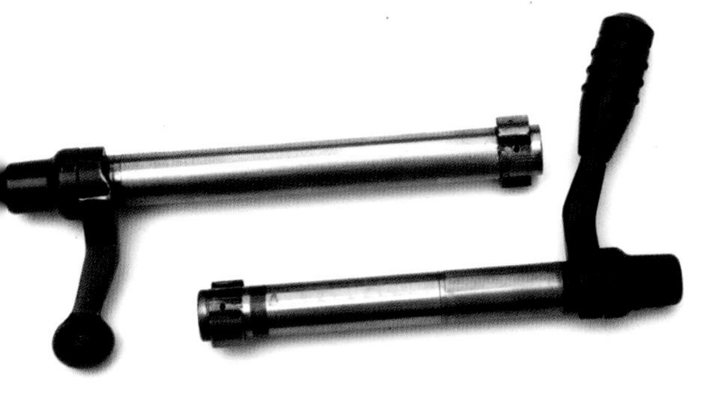

Steve fits collars or plates and then grinds the bolt diameter to make a perfectly concentric and minimal fit between the bolt and action.

The bolt face is trued for a perfectly concentric fit to the action.

Bolt lugs are trued and lapped to the action abutments to ensure all the surface areas are perfectly in contact.

bolt to enlarge the outside diameter for a tighter fit. These sleeves can be made in varying ways: half sleeves, total circumference or bolt plating. The sleeves are usually made from high grade 316 or 420 stainless steel. EN16T or EN24T steel is used if the bolt is to be blued. Either nickel or hard chrome is used for plating. In the case of both plating and sleeves, the surface is reground for a 0.001in fit in the action.

The focus now turns to the bolt face. On Steve Bowers's Remington 700 actions the face is machined so that the bolt face shroud sits into the back of the barrel/chamber. This is a 0.001 thou clearance fit and supports the bolt, making it concentric to the barrel.

Lap the bolt lugs, using grinding paste, or cut true to the abutments in the receiver ring.

At this stage the extractor can be converted to a Sako type. This is a small change that makes a big difference by greatly improving the force on the case's rim for positive extraction and reliability.

There are many actions that use an incorrect face to the cocking piece. Cocking the bolt causes it to rise slightly and this creates unwanted upward pressure that can lead to inaccuracies. Sleeving helps, as does an angled contact between cocking piece and trigger sear engagement, allowing it to drop downward as the trigger is pulled.

Sleeving the rear of the bolt helps keep the trigger sear perfectly engaged when cocked.

The thread on the barrel tenon is cut so that it matches the recut action thread. Everything should now be perfectly concentric and the harmonics on firing will be in tune with each other. The incredible variety of thread profiles available means that this is a hotly debated aspect of custom gunsmithing.

TOTAL ACTION SLEEVING

Sleeving usually involves an aluminium, steel or titanium sleeve that fits over the donor action. The action is slid in and then epoxy glued to form a strong union that gives improved rigidity. If the sleeve extends beyond the receiver ring heavier barrel profiles can be fitted, as they are supported better by the sleeve. There is also an increased surface to bed the action. Aluminium bedding pillars complete the transformation. This technique was most popular in the 1960s, but it still has its merits and could be used for a retro custom gun project. Davidson and Dorsey sleeves use a Remington M700 action or clone as the donor item and the rigidity is transformed in a CNC aluminium sleeve that also gives a better footprint for bedding into a stock and an integral recoil lug. Typically a sleeved Remington action increases rigidity from 0.5×10^6 action to 3.3×10^6. Full blown custom actions, however, can achieve 10×10^6 lb.

CUSTOM ACTIONS

In my experience this is where the good transcends to the excellent. A factory action will always be a factory action, however good it is or trued it has become. With a custom action designed from the ground up you have a no-holds-barred approach providing an uncompromising action for the best, most expensive project. As it is such a pivotal part of the custom rifle, you should start here if you can afford it. Otherwise you may later find yourself looking at your new rifle and still wondering about what it might have become, and that's not what a custom rifle is about.

Custom actions are usually designed as a single action, or series of actions, to accommodate differing cartridge sizes. They are sold without a barrel, magazine or scope mounts. Some custom action manufacturers now offer complete rifle builds, but most concentrate on producing an action and its associated ancillaries to the highest specification possible. Let's look at the various custom actions available today and some of their common features.

Materials

Stainless steel still remains very popular among both custom riflemakers and customers. Usually a 416 grade is used that provides strength and rigidity, resists corrosion and erosion, and can

Custom actions are made to precise tolerances and need no truing up, but they are expensive.

With a custom action the orientation of the bolt and loading port can be specified to suit your own shooting styles and preferences.

be finished in various ways as desired, including a pleasing polish, brushed stainless, mopped or even blacked or painted. Chrome moly is still used and makes a traditional action for those wanting a blued finish, while still achieving all the advantages of high quality and close tolerances that a custom action provides. Aluminium offers lightness and so a large action can be produced for a bigger bedding area without being too heavy. All wearing parts and bearing surfaces are made from heat treated steel for longevity. Titanium is both light and strong but hellishly expensive. Some manufacturers use more than one metal type in the action for differing wear resistance or looks, such as aluminium/titanium or steel/aluminium.

Orientation

This refers to not just whether the action is right- or left-handed. With a custom action you have the option of placing the loading port on the opposite side to the bolt handle. If most of your shooting is from a bench or out varminting off a bipod, it is very handy to operate the bolt with your right hand and then load a round in the left-side port with your left hand, instead of releasing the grip with your right hand to pick up a new round and load in the right-side port. This is obviously reversed for left-handed shooters, although any orientation is possible: several guises of the old Stolle Panda action had ports on both sides and many new actions can be ordered as such. Most manufacturers of custom actions denote their actions as RBLP (right bolt left port) or LBRP (left bolt right port), for example.

Lock Time

One of the most important aspects of any firing cycle is the speed at which the firing pin hits the case's primer for ignition after you have squeezed the trigger. With a faster lock time it is less likely that any variations in hold, movement and vibrations will take place to disrupt the accuracy. A Remington 700 action has a typ-

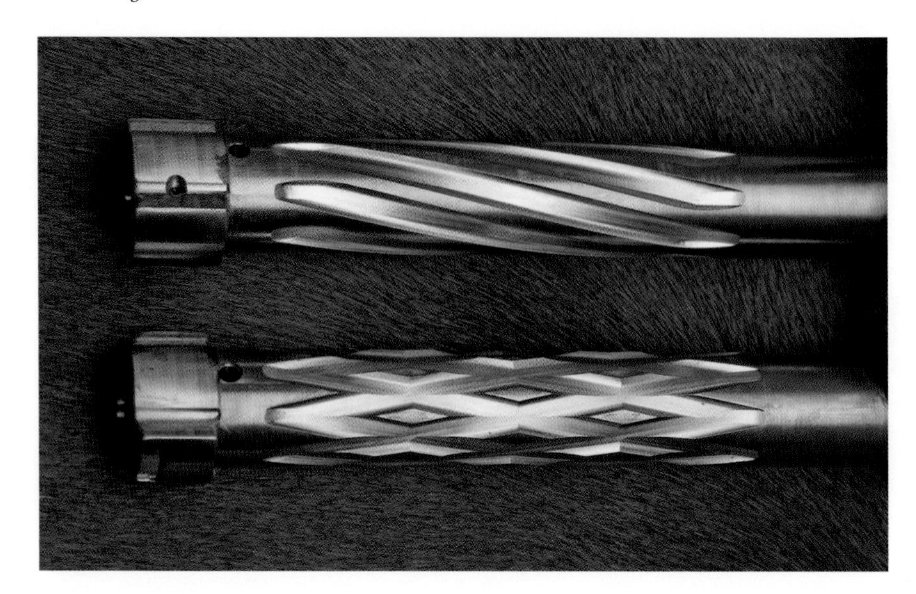

Custom actions can come with various bolt styles. Fluting a bolt, here seen with spiral and diamond fluting, is a good way to show individuality in a custom rifle.

ical lock time of 2.9 milliseconds and the Remington 788 achieves 2.4 milliseconds. With a full-blown custom action, however, you should expect 2.0 milliseconds. Altering components such as springs and firing pins in a stock action can reduce lock time. Custom actions come with these features as standard, but you can go further by upgrading to titanium firing pins for lightness, fluted firing pins to increase speed and specially profiled springs for maximum tension release.

Fluted Bolt

This is both a cosmetic and a functional addition to any custom action. Custom makers often have their own instantly recognizable style in fluting, perhaps a simple straight flute along the bolt body, spiral fluting, interrupted flutes or even spiral flutes crossing both ways, which results in a diamond effect. Blacking the inner surfaces of the flutes against a polished bolt body gives a pleasing effect. Fluting a bolt helps to reduce the weight and limit the chances of the bolt galling in operation as there is less surface area around its circumference to which it can bind. If dirt or grit gets into the action it can collect in the flutes, helping to prevent the bolt from locking up until you remove them (although I have known water freeze in the flutes and lock a bolt solid).

Coned Bolt

A coned bolt face has two advantages: safety and function. If a primer is pierced the escaping gases are then angled away into the bolt's exhaust ports. The functional advantage is that the bolt can sit further into the barrel or chamber for better support of the case.

Bedding Surface

The underside of the action is a key part of the overall bedding surface and as such will directly reflect how well an action sits in the stock. Although rounded action bottoms are popular, for example on the Remington M700 and its many clones, flat bases as found on some BAT, Stiller or Stolle actions are easier to bed. Regardless of the shape, it is crucial that the surface is concentric or perfectly flat, and not twisted in any way, as any imperfections may potentially result in less accuracy.

Recoil Lug

This can be either separate or integral. Separate lugs are perfectly ground to be flat and increase the surface area to bed into the stock, also ensuring a non-tensioned fit between barrel and action face. They are often pinned to stop rotational torque (sometimes with two

Recoil lugs are designed to spread recoil but also help to bed the action to the stock. The lugs may be either integral or separate, as here; both types work well.

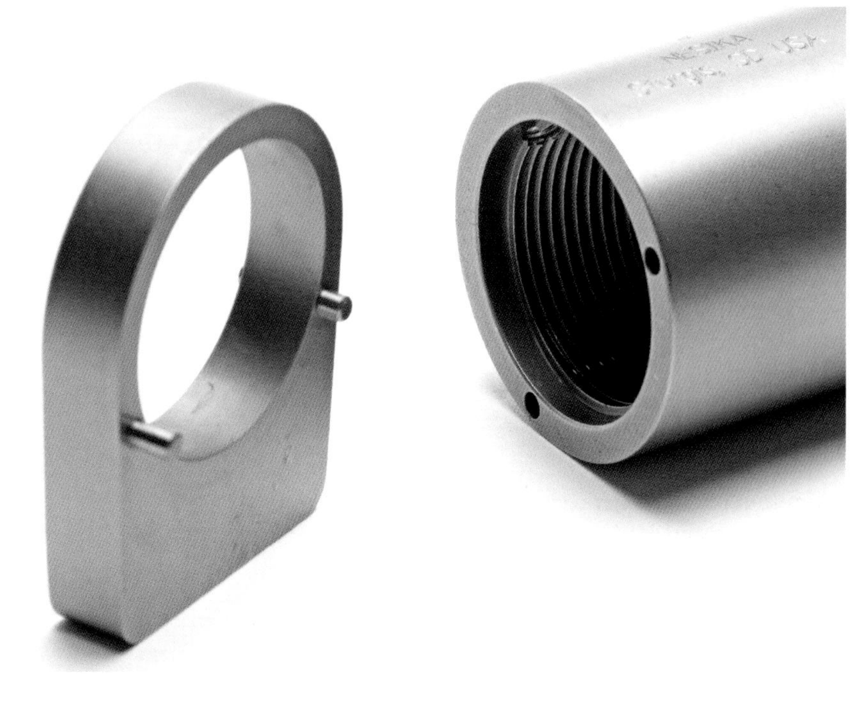

pins) and much thicker than the original factory lug, again for better rigidity. Integral lugs can be located either up front on the face of the action, protruding just beneath the barrel or action joint, or they can be farther back along the action's bottom to create a more balanced bedding section, sometimes offering a position for three stock securing screws. These can be integral and machined as part of the whole action or located via a dovetail.

Integral Scope Mounting

With a custom action it is possible to upgrade the scope mounting system. Although dovetails are still used by benchrest shooters, the use of an integral Picatinny rail allows a more universal scope mounting fixture. Often the extra positioning of these rails extends the whole action length, allowing even the largest scope to be mounted easily. They also have the advantage that they can be machined with a Minute of Angle (MOA) bias. Being able to tilt the scope down at varying degrees of arc from 10 to 50 MOA gives shooters further opportunities to make adjustments for extended long-range shooting.

CUSTOM ACTION MAKERS

This is where your dream rifle really begins. There are some very talented machinists out there with the abilities needed to machine a classic Mauser type action from a steel billet in the traditional way right through to CNC machined actions of a unique design that make quality and close tolerance parts a reality.

A selection of stainless steel custom actions: (bottom and left) John Carr actions; (top) DCR action.

Stiller Precision Firearms

Stiller Precision Firearms (SPF) uses only the best state-of-the-art CNC machines and about twenty lathes and mills, as well as EDM machines to cut out bolt raceways and centres in the action. The number of machines means that each can be dedicated to a particular action, thus keeping the process to very tight manufacturing tolerances. SPF makes many differing types of custom actions, using aluminium, stainless steel or traditional chrome moly.

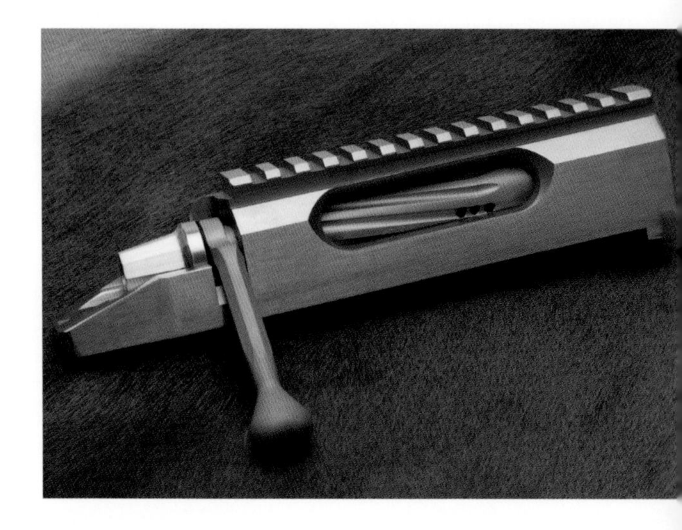

Stiller Precision Viper SS action with fluted bolt, extended straight bolt handle, flat bedding surface and 20MOA scope rail.

Aluminium

Aluminium actions come from the Viper, Rattler, Cobra and Python series. These actions all use a one-piece aluminium body with barrel inserts of 416R stainless steel hardened to 41 RC. All aluminium parts are hard anodized for wear resistance and to ensure a smooth cycling action. They are available as RBLP or LBRP configurations and a black iron nitride coating is applied to reduce galling (binding), which is common where aluminium surfaces come into contact. All bolt lugs – two on these models – are lapped at the factory. The bolt face is a coned head with a M16 style extractor and the bolt body is made of 4140 hardened steel. Additional features include a spiral fluted bolt finish with a chamfered ejection port and the trigger hanger is designed to fit a Jewell or other Remington-style aftermarket trigger. An attractive feature is that the integral scope dovetails to the action top. SPF actions also come with scope rings included as standard. The dovetails are compatible with any Davidson-style ring or Kelbly rings.

Stainless Steel

Stainless steel models come in competition (Diamondback, Viper SS) or hunter models (Predator). The competition models are all 416R or 17-4 ph stainless steel hardened to 41 RC with RBLP or LBRP configurations. The bolt hole and raceways are drilled and electro-discharge machined to match the rails and bolt body/lugs support with exact tolerances. The finish is satin brushed. The bolt, which is treated with black iron nitride for a smooth non-galling action, is made of 4140 steel hardened to 40 RC and the locking lugs are lapped with a 30-degree coned bolt and M16 extractor. Davidson mounts or a Picatinny rail are available in standard and 20 MOA offset, while the larger Viper SS has a built-in Picatinny rail with 20 MOA offset and built-in recoil lug. This big Viper SS is a very good candidate for building a long-range custom varmint rifle.

From a sporting point of view I really like the stainless steel Predator action. This is a direct clone of a Remington 700 or 40X action and thus all aftermarket triggers, stocks, mounts and similar items will fit. It is, however, nothing like the Remington in terms of quality: it is far superior in every way as tolerances run with a bolt fit to body typically of 0.004 to 0.006 thou. The Predator is available in short and long action lengths, as a single shot or a repeater with 223, 308 and magnum bolt faces. There is even the PredXtreme model, which is designed for the 338 Lapua magnum or 378 Weatherby based cartridges. All action bodies use 416R stainless steel hardened to 41 RC and they share the same black nitride bolt, factory lapped with spiral flutes and incorporating an anti-bind rail for smooth operation, especially when ejecting a spent case. The Predator repeaters have an extended magazine cut-out for use with HS Preci-

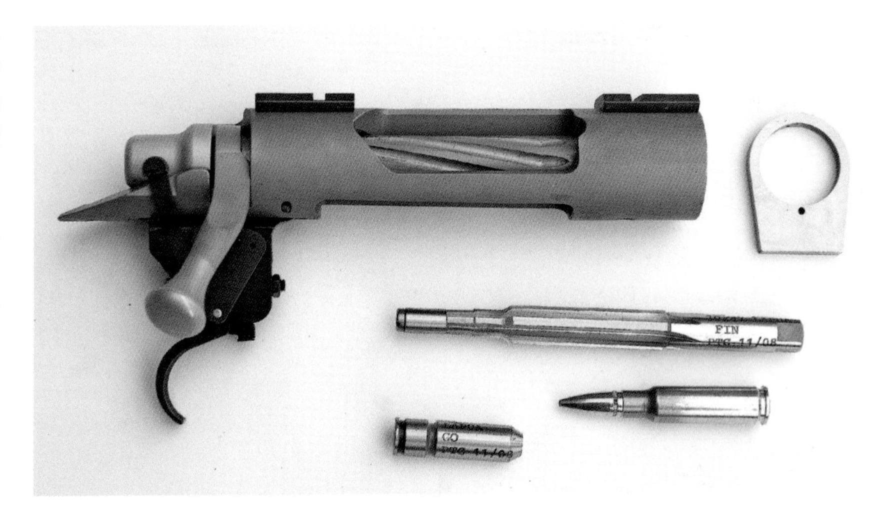

Stiller Predator stainless steel action, which is like an improved custom-trued Remington 700. It is shown here before my . 30-47L project with Steve Bowers.

sion or similar extended magazines for factory Remingtons. Scopes can be mounted in several ways using a Davidson dovetail, a Picatinny rail or standard front Remington bases, which fit both front and rear mounts. The action profile has a full rounded outside dimension and a small profiled ejection port actually means a more rigid action. A standard Remington-style firing pin with 0.068in diameter is used, but aftermarket fast lock time models can be fitted, if desired.

Classic Chrome Moly

The Classic line is designed for hunting use with chrome moly actions. It is a 4140 chrome moly steel 'drop-in' replacement for the Remington 700 series actions, with the same bolt faces as the Predator but without the extended magazine well. The .223 bolt face uses the Sako style extractor, while the other sizes use a modified M16 type, although all are really aimed at the classic blued gun market and come in white for the finish of your choice. I fully agree with SPF's claim, 'Why buy a Remington, spend hundreds of dollars blueprinting and still not get the quality of this action for the same money?'

BAT Actions

BAT actions are the brainchild of Bruce Thom of the BAT Machine Company, which was founded in 1990. His first design was a falling block action, but Thom is now hugely regarded for his precision benchrest, hunting and tactical bolt actions. BAT actions are very popular in Britain and are used by many top custom gunsmiths in their projects. Styles include pure benchrest, sporter, hunter, tactical and a choice of two- or three-lug locking systems. Actions are made from 17-4 stainless steel or 4140 chrome moly and pre-hardened to 36-38 RC. The bolts are 4140 chrome moly to run smoothly in the stainless action body. The bolt handle is stainless and is welded on, except on the EX model and the three-lug actions, where they are screwed in. All two-lug actions have a bolt lug diameter of 0.719in with an action bore diameter of 0.720in and bolt diameter of 0.7185in, giving a clearance of 0.0015in. Ther are similarly tight tolerances in the three-bolt lug actions, a full 0.900in diameter lug arrangement with a bore of 0.900in, bolt diameter of 0.8985in and 0.0025in clearance. Close tolerances mean a very precise action. The lock time, too, is very fast at 2.2 milliseconds with a short firing travel of 0.210–0.230in.

Choosing an action from BAT is easy as there is a huge selection of action types to suit short- or long-range benchrest, F Class, Palma, varminting, tactical, long-range varminting and .50 cal. The small actions are the S and SV, which handle .22 Hornet to PPC and BR case sizes, whereas the B and 3L actions are designed for .223 to .308 cases; the DS is only for PPC and BR cartridges. The S is the same length as

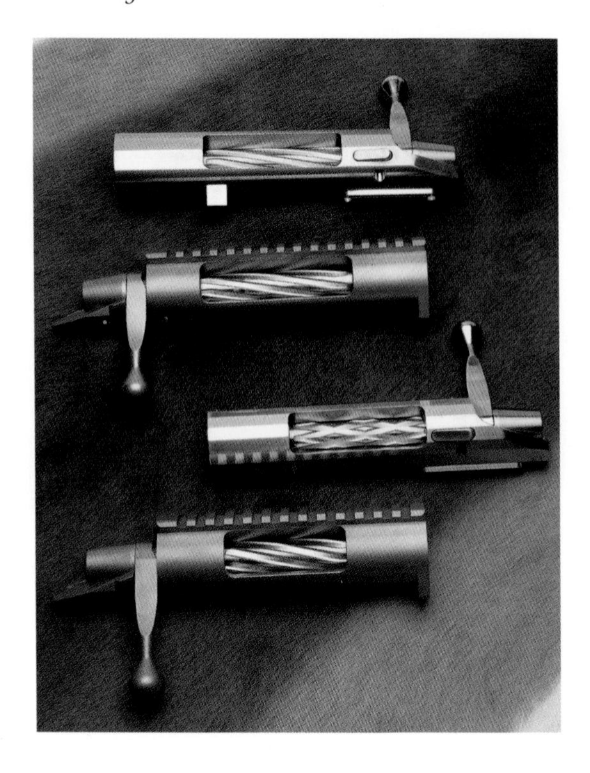

BAT actions are considered among the best custom actions available, blending precision engineering with good looks and made to suit all cartridge types: (top to bottom) Model M, Model HR, Model B and VR model.

the SV but is available in more options. It is 6.0in long and can be stainless steel or chrome moly construction with a weight of 32oz. The B differs in that it is 7.6in long with a larger port and 33oz weight. The M model works with .243 to .338 Lap magazines, while the L model can handle up to 408 Cheytac. The 3L and 3LL are both three-lug bolt lug systems that have the advantage of a 60-degree bolt lift compared to the 90-degree lift from the two-lug bolts. The Model L is 10.0in long and thus has a longer barrel tenon for heavier barrels of 1.45in. There is a faceted stainless action body and separate recoil lug positioned below the reloading port. The weight starts at 80oz and the long bolt handle has a teardrop bolt knob. The VR model is a Remington Model 700 clone, weighs 34oz and has a Remington short action footprint with an 8in length and 1.35in diameter. It is a repeater action, so will accept any Model 700 bottom metal, but can be ordered as a single shot if desired. The recoil lug is integral and the VRPIC version comes with a one-piece Picatinny rail, zero or 20 MOA. The HR is a Remington long action clone suited for larger calibre cartridges.

The biggest action is the EX model, which is designed for ranges over 1000 yards or 50 BMG shooting. It is 12in long and is available with two

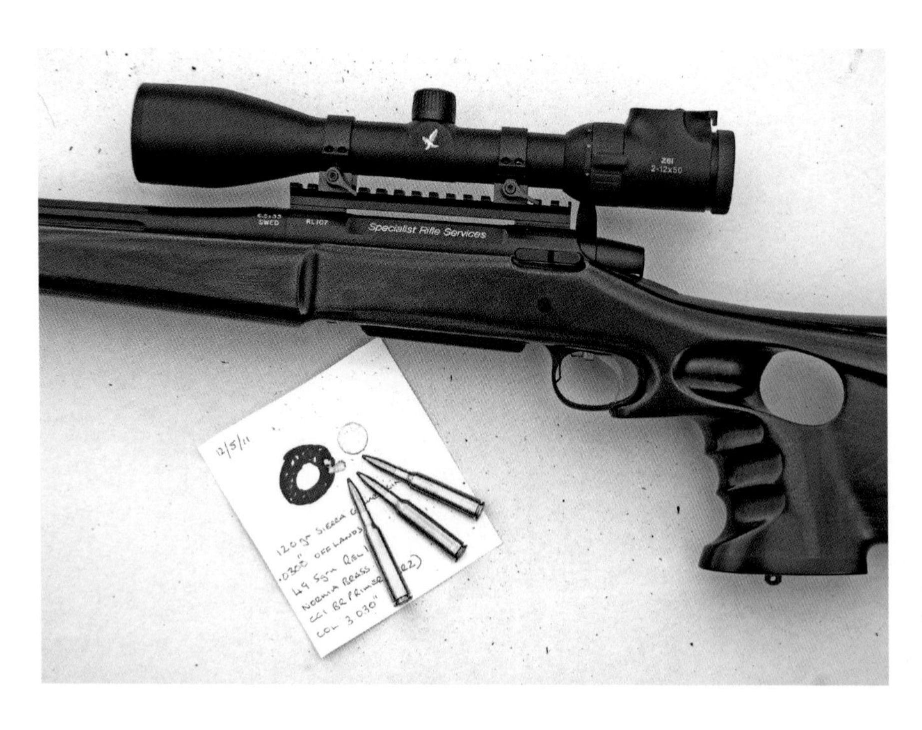

Steve Bowers custom BAT HR in 6.5 × 55mm and custom laminate stock.

exterior action shapes, of which the rounded looks very clean and weighs 128oz. Barrel diameters up to 1.75in can be accommodated with options throughout the range that include actions with dual ports, as well as the usual right/left port and bolt arrangements with the addition of a bottom port also. Recoil lugs can be integral or separate, if you wish, and the body shape on most actions can be round or octagonal. The bolts can also be non-fluted or spiral, straight or diamond fluted. Extractors may be a Sako type or sliding plate arrangement. It would be impossible here to list all the options and variations between each individual action design, but there is almost certain to be a combination that suits everyone.

Stolle Actions

The popular Panda action, designed by Ralph W. Stolle and made by Kelbly's Inc, was the first to be manufactured from aluminium. This action is a good choice both for benchrest competition and for sporting use where the owner wants a precise rigid action capable of developing tiny groups on paper.

Panda

The Panda is available in either standard length or 0.850in longer with an integral recoil lug. There is a good selection of bolt and port orientations as well as short actions for a small bolt travel, which allows for fast bolt manipulation. Standard features include an aluminium bolt plug and spiral fluted bolt, an enlarged-profile bolt knob and an aluminium trigger guard with screws included. The Panda action weighs only 30.5oz and bolt faces can be ordered in .222, PPC, 308 and magnum sizes. The bolt face is coned or flat and can have a right or left orientation, with a single port either left or right or a double port on both sides. A dovetailed scope rail integral to the top of the action accepts Davidson scope rings, which are popular with benchrest shooters.

Atlas

A more conventional hunter class action is the Atlas model, which is a Remington 700 clone but is made from 416RS steel and weighs 29.5oz. The bolt is forged from 4140 alloy steel. The action is strong and rigid in construction with very close tolerances to all surfaces. It is a rounded action with the rear of the receiver on a level plane with the front base, unlike a factory Remington 700, which makes it easier to align scopes. A two-pin recoil lug on all Atlas actions makes it less likely to rotate or bind if the action is set up as a switch barrel system.

Borden Actions

Borden Rifles/Borden Accuracy is a family-run business founded by Jim Borden in Springville, Pennsylvania. The British agent is Callum Ferguson. A Borden is probably the best-made custom action money can buy and represents perfection at the heart of any custom rifle project.

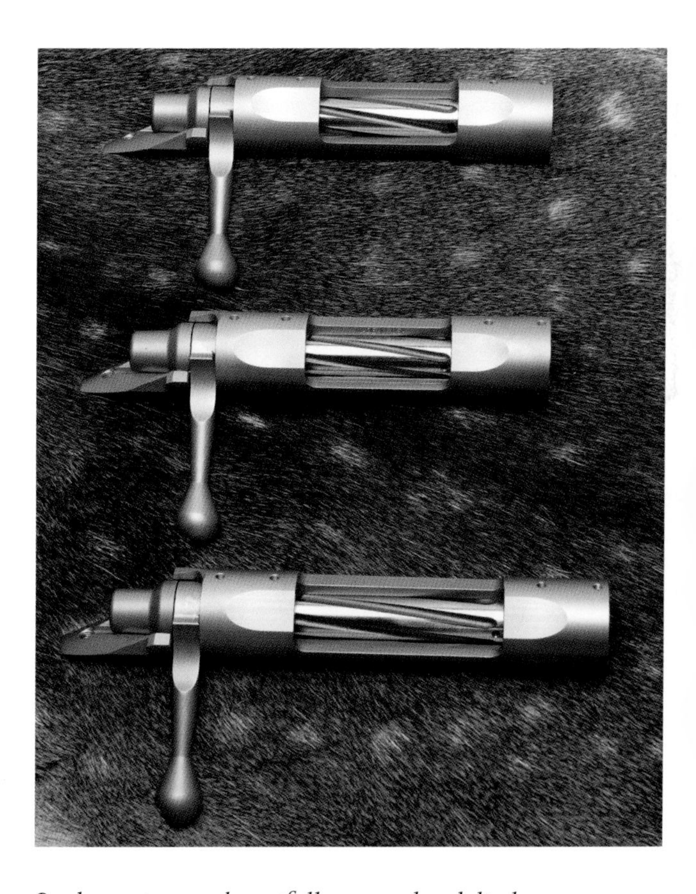

Borden actions are beautifully executed and display real precision, with only two-thousandths of an inch clearance between bolt and action: (top to bottom) Alpine single shot, Alpine repeater and Timberline.

Callum Ferguson full custom Borden actioned rifle in repeater trim, bedded into a McMillan stock with a Krieger barrel.

The Alpine and Timberline actions are made specifically to be used in stocks inletted for Remington 700 actions. The tight bolt-to-receiver fit of these actions (less than 0.002in) makes them ideal for precision hunting rifles as well as target use. They are manufactured to true benchrest tolerances and yet are reliable in the field. The Alpine actions fit stocks inletted for Remington 700 short actions (repeater action weighs 26½oz) and the Timberline actions fit stocks inletted for Remington 700 Long actions (repeater action weighs 28½oz). Both are available in right- or left-hand orientation. Key features of the Borden action are their hardened (40 RC) stainless steel action body with hardened (44 RC) and ground 4140 bolt body with a clearance between the two of less than 0.002in and a superb 0.0002in tolerance for a perpendicular bolt face. There is no anti-bind rail as the bolt lugs are designed to glide perfectly within the action's raceways. The action is further enhanced by a spiral-fluted bolt body and silver-brazed bolt handle, with an optional double-pinned recoil lug. Standard and magnum port lengths are also available for each model.

Four bolt face sizes are available in both single-shot and repeater models: Dakota, Magnum, 308 or 223. The 308 bolt face extractor is able to extract and eject PPC case size cartridges. Special feed rails are designed to reliably feed even the short magnum rounds as well as ultra magnum length cartridges. Sako-style extractors are fitted, although M16 type extractors can be installed on .308 and magnum bolt faces if desired.

Nesika Actions

Nesika's slogan is 'five shots one hole'. The firm was one of the first action makers and its products were very fashionable when I became interested in custom rifles in the 1990. It offers a complete line of hunting, long-range and tactical actions as well as semi-custom rifles. They are machined from aircraft grade 15-5 stainless steel to tolerances of [plus minus] 0.002in. The one-piece bolts are made of 4340 chrome moly with the addition of Borden Bumps, sited behind the locking lugs and just in front of the bolt handle, to improve consistency and precise

Nesika actions were among the first to be imported for custom rifles. Dolphin Gun Company supplies an impressive range.

lock-up of the bolt within the action to within 0.001in. Bolts can be ordered with spiral, straight or non-fluted design to suit your rifle's overall look. There are four models, Classic, Round, Hunter or Tactical, catering for any custom rifle preference.

Classic

The Classic action is more of a classic benchrest design than hunter orientated, with strong faceted side walls and large bedding area. The three types available – Model R, E and G – differ only in their length to accommodate differing cartridge lengths. The bolt diameter is 0.700in, so not magnum size and ideal for medium cases.

Round Model Actions

These are ideal for match-grade accuracy and all-round versatility. The 0.750in diameter bolt is well suited to Lazzeroni and Lapua sized magnum cartridges. The model designations – Model J, K, L, S and M – relate to the differing outside diameters and bolt sizes as well as the overall length. The Model J is good for standard cartridges: it is 7.36in long with a 1.35in or 1.47in diameter action and a 0.700in bolt. The K and L models are longer and have an additional option of a 1.70in diameter action for 30-06 class cartridges. S and M action sizes range from 8.11in to 8.86in length, respectively, with only 1.35in or 1.47in diameters, but with the larger 0.750in bolt diameters for magnum calibres.

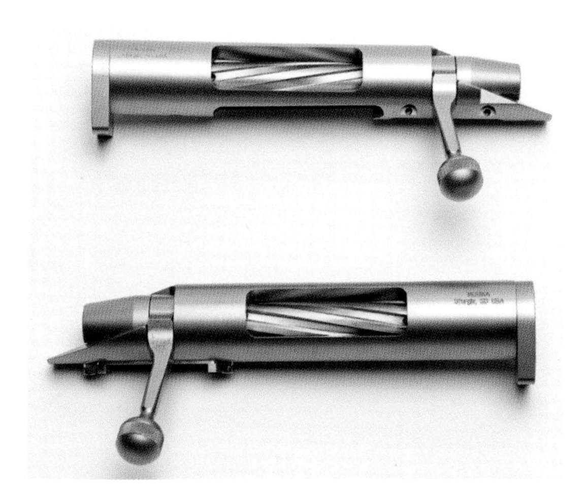

The Nesika Model K action is available in right- or left-handle configurations.

Hunter

This lightweight action has a larger ejection port for reliability in the field and faceted side walls to the action sides. It is a lovely looking action and eminently suited to the highest quality custom rifles. Models are again aligned to cartridge type and length. The Model C, T and V have the same action diameter of 1.35in and 0.700in bolt diameter, but are 7.36in, 8.11in and 8.86in long, respectively, for increasing sizes of cartridge capacity. The Model MH is a large action at 8.86in long and 1.35in outside diameter, and it has a larger 0.750in bolt for magnums.

Tactical Model Actions

These have action lengths to fit a variety of calibres. Features include a fluted bolt and a 1913 Military Standard scope rail for universal scope mounting solutions. Depending on whether standard or magnum cartridges are to be used, the lengths are 8.11in or 8.86in, with action diameters of either 1.35in or 1.47in and bolt diameters of 0.700in or 0.750in. There is a double-pinned, separate recoil lug, long tactical-styled bolt knob for fast handling and easy grip, and the bolt rear shroud can be either stainless steel or aluminium.

Geske Actions

Gerry Geske is a Canadian enthusiast for long-range hunting who now lives in Montana. He has designed an action for magnum calibres for long-range use. Gerry uses stainless steel 17-4 for his single-shot receiver. The action begins as a $1^{9}/_{16}$in diameter bar about 10in long, which is drilled centrally for the bolt hole and then the action is trued around it. It is available with both left- and right-hand bolts and ports. It is a big action, 1.55in in diameter and 9.25in long, with a large loading port of 3.60in for magnum cases. It has an overall weight of 52.5oz and

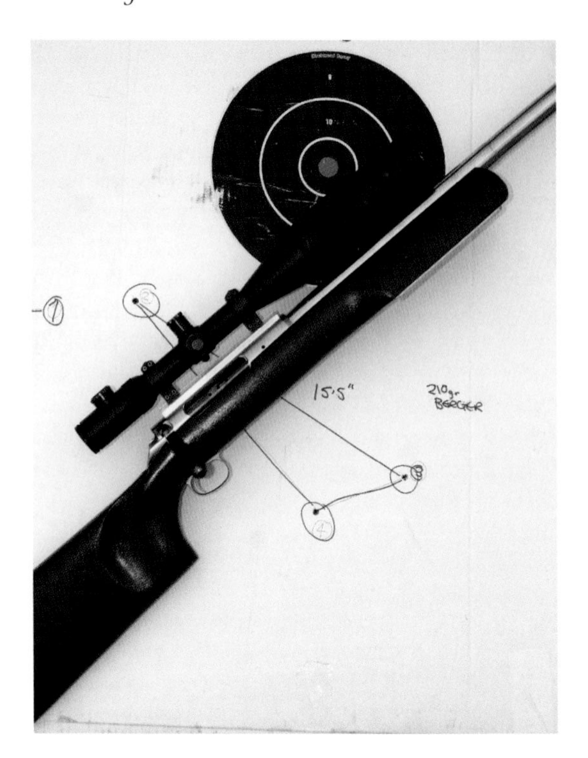

Geske action chambered in 300 Remington Ultra Magnum and Lee Six stock, seen here set up as a rail gun. The group was shot at a distance of 1500 yards.

can handle cartridges up to 4.2in long. There is no integral scope mount, but it is drilled and tapped for separate bases. The trigger hanger will accept Remington style units, usually a Jewell or Timney.

The bolt's diameter is 0.890in, which is the same as that of the three lugs, providing more strength and half as much lug shear stress on firing as a Remington M700 action. The stability of a cartridge in the chamber is also improved by the three-lug system, muting harmful vibrations, and there is only a 65-degree opening angle on the bolt. The wide bolt has twin cocking cams that provide an effortless and smooth cycling action. Another interesting feature is that the abutments, which are usually machined directly into the action wall, are replaced in Geske actions by a separate, threaded lug insert that perfectly matches the locking lugs. Lock time, about 2.1 milliseconds, is very fast due to the stout firing pin spring and the light titanium firing pin itself. I have been using a Geske action

Barnard actions are popular with long-range shooters in Britain. The GP and S actions are capable of record-breaking accuracy.

in a 300 Remington Ultra Mag, built by Norman Clark, to great effect as an ultra long-range varmint and steel target gun.

Barnard Precision

Made in New Zealand, this three-lugged bolt action is a firm favourite in Britain. For more than twenty years Barnard Precision has supplied variants of the Model P and Remington-style actions to suit any custom project .

Barnard Model P Action

The original design appeared in 1982 and was intended for Palma or long-range shooting, but it has since been used in many long-range hunter in the UK. It is a single-shot cylindrical action, in right- or left-handed configuration. The PC action can be ordered with a bolt and port orientation to suit your preference. The bolt has three large forward-locking lugs with Sako style extractor. It will accept a case up to a magnum-size rim diameter of 0.534in and is hand-lapped to the action. This is made from 4340 chrome/moly/nickel steel hardened to 38 RC and machined after hardening for concentricity. Scope mounting is via separate bases, but Anschütz standard dovetails can be integrally cut, if preferred.

Barnard Model S and SM Actions

These actions are cylindrical, like the Remington-type action, and have the same outside diameter, but they employ the Barnard three-lug configuration. This means that the bolt lifting arc is shorter than on a standard Remington 700, the ejection/loading port is different and the action is 10mm longer. The action weighs 35.06oz and is made of 4340 chrome/moly/nickel steel hardened to 38 RC with the bolt hardened to EN39b, ground and machined after hardening to ensure straightness and concentricity. The bolt lugs are lapped to the action abutments and a Sako-type extractor is equipped with a bolt face plunger-style ejector. The actions will take Remington aftermarket trigger assemblies and, despite the size difference, they can be fitted to the numerous stocks designed for Remington 700 actions with a little custom work. The SM repeating action is a derivative of the Model S, with the difference that it is designed to accept a five- or ten-round HS Precision magazine and floor furniture/bottom metal.

Barnard Model GP Action

The GP is the largest Barnard action and designed for large calibres. It follows the conventional cylindrical, forward-locking turn bolt design and is available in either left- or right-handed configuration in single-shot only. It is 12.0in long with an outside diameter of 1.97in, and weighs 125oz. Again the bolt has three forward locking lugs and is equipped with a Sako-style extractor and bolt face, which can accept any case rim up to 0.804. The action is made from 4340 chrome/moly/nickel steel with a solid bottom cylinder and a narrow loading port for rigidity. The action is supplied with a 6061 aluminium bedding block (12mm recoil lug optional) with two parallel points of contact to the action and an anodized 6061 aluminium Picatinny-style scope rail (30 minutes of angle), and modified two-stage Barnard trigger. Cartridges up to 5.50in length can be cycled through this action.

Dolphin Gun Company

The Dolphin CST DGC is an action built in-house to the company's own design. It is as good as it gets and has many innovative features. The action has been engineered from the ground up to provide the most accurate shooting platform obtainable. It is fitted with a Cone-Lok[TM] bolt design with cone-shaped locking lugs. A cone-shaped floating bolt handle ensures the rear and front of the bolt are seated centrally and square when loaded. This also produces continuous and uniform pressure of all the locking lugs to maximize the accuracy potential. An innovative feature is the zero-lift cocking piece, which has

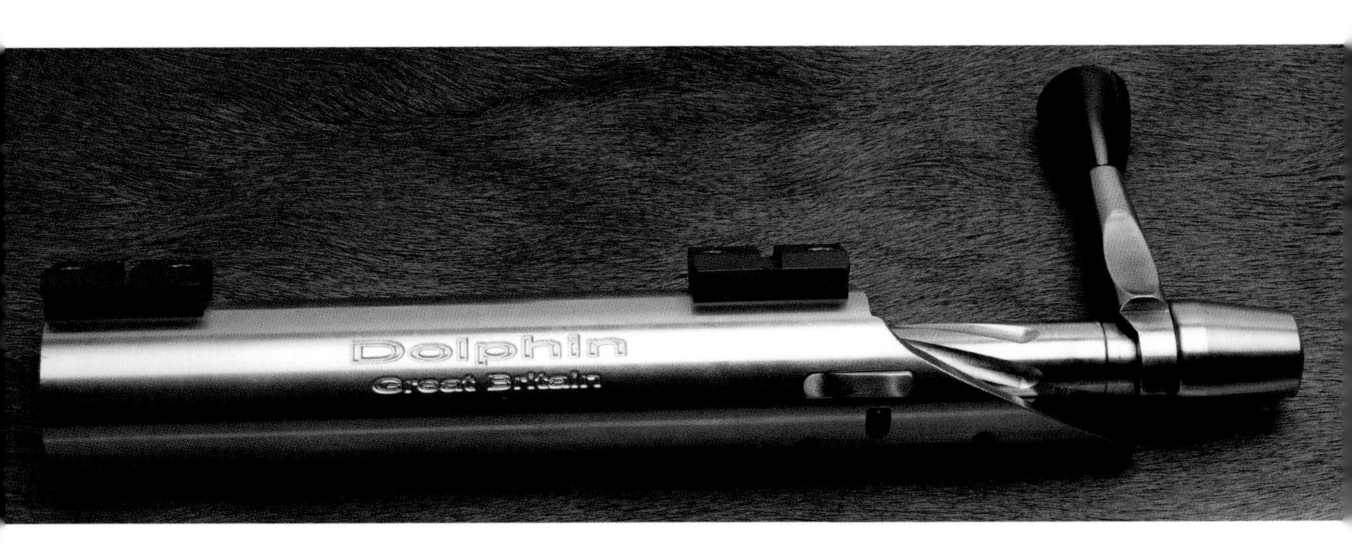

The CST DGC action from Dolphin Gun Company is probably the best designed action ever made.

The Cone-Lok[TM] bolt design, with cone-shaped locking lugs and a cone-shaped floating bolt handle, maintains a perfectly concentric bolt when loaded for extreme accuracy.

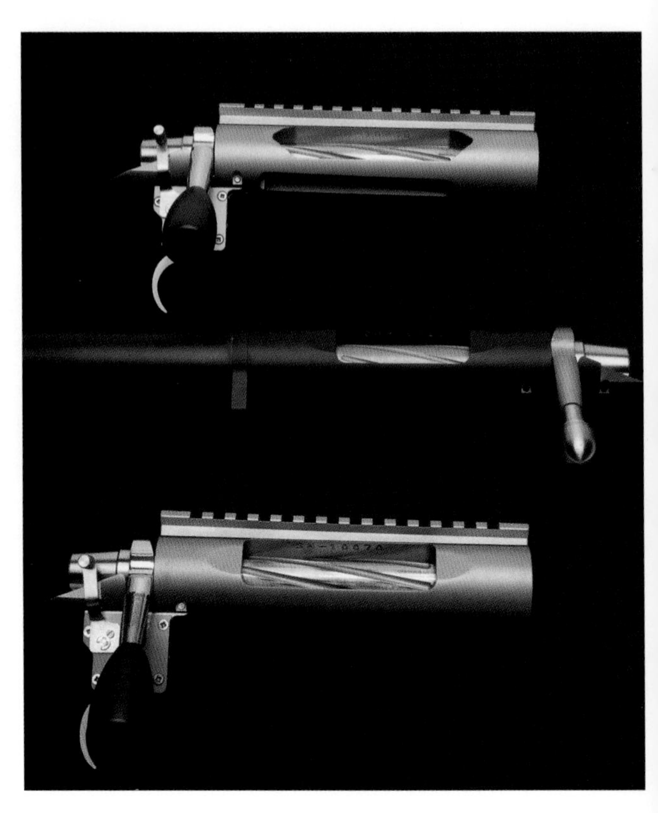

The recently introduced DCR actions provide a range of Remington clone fitment-type, precision-engineered actions for all orientations and cartridge sizes.

flanges extending from the sides of the cocking piece to stop upward lift of the firing mechanism during bolt lock-up, eliminating an angled sear engagement.

The three-lug design with 60-degree bolt throw and magnetized 0.062in firing pin tip maintains contact with the firing mechanism while allowing a perfectly centred position in the bolt for positive ignition. All actions come with magnum and 308 bolt bodies to make it easy to change barrels for the different calibres.

Devon Custom Rifles

Devon Custom Rifles (DCR) is a new English action maker based in Devon. The owner, Gary Alden, uses only CNC milling, grinding machines and lathes to produce home-grown actions as well as high-quality ancillary parts and complete custom rifles. John Davies joined DCR in 2008 to produce custom actions and rifles in response to the sporadic availability of actions from America. All actions and parts are designed in-house, so clients can have their say about creating an action and rifle exactly to their own specification. The DCR action is based around the venerable Remington Model 700 action and so can utilize all the aftermarket products available. DCR also manufactures custom bolts, firing pins, bottom metal/trigger guards, magazines, scope rails and lugs. DCR has recently introduced a precision replacement trigger assembly for the Remington model.

GBR England

Custom actions made by GBR are similarly based around a modified Remington 700 model that has been made to much finer tolerances and with a number of differences. The action is circular in design and is made from 4140 chrome molybdenum steel with a diameter of 1.35in and 7.85in long. It is also hardened to 37-40 RC and weighs in at 450

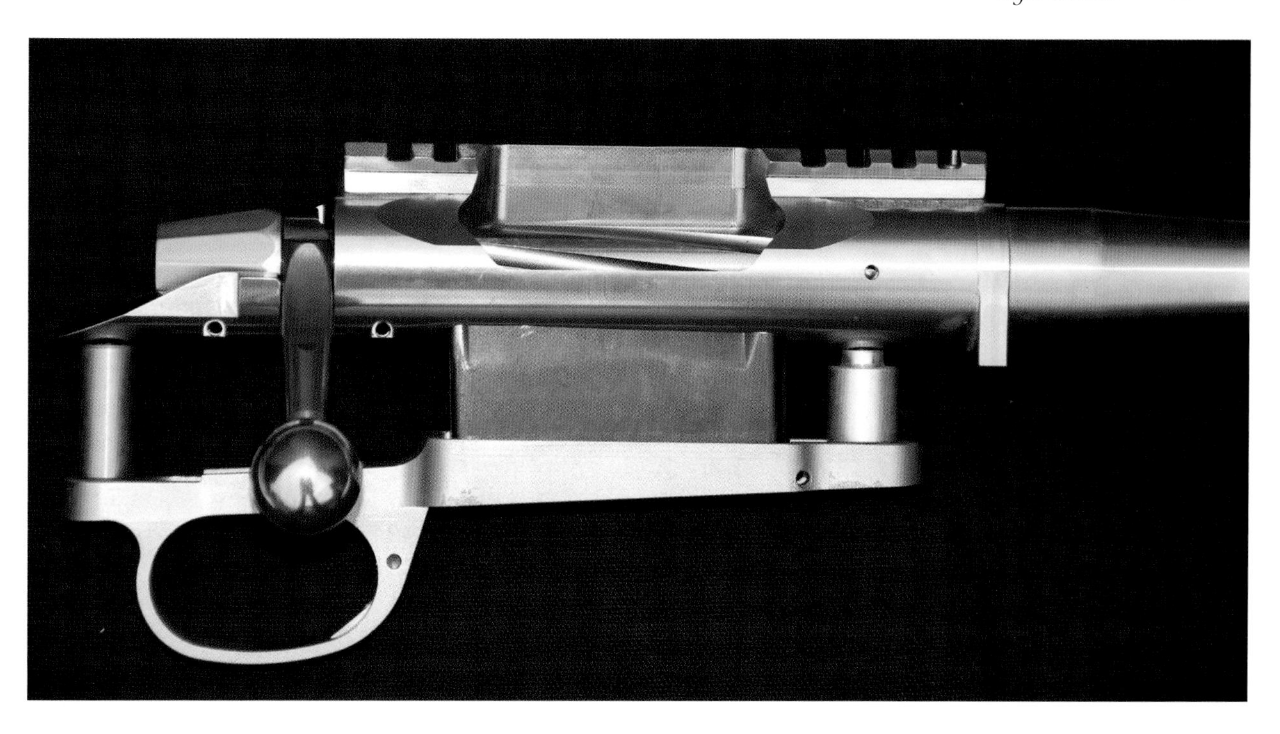

John Carr actions are made primarily for the trade. Individual custom rifle makers can order their own take on a Carr action for very personal results.

grams. The bridge section is machined at the horizontal axis, so there is no step down as on original 700 actions. You can now fit the same scope bases or opt for a GBR one-piece 4140 chrome moly steel Picatinny rail mounted via M4 threads. This can have zero MOA bias or 20 MOA on request. The action has a smaller ejection port of 2.4in that improves the action's stiffness, the 0.275in thick recoil lug is ground flat and then it is pinned via a 2.3mm dowel to the action face, made from the same steel as the action. The bolt is still a two-lug type made from 416 stainless steel, while the action is machined from 4140 chrome moly steel. This allows a blued finish to be applied, making it suitable for traditional custom rifle builds benefiting from surface beauty and inner precision from the close tolerance manufacture. The bolt body of 0.7in diameter has a helical fluted surface. A Sako extractor is fitted and the bolt release mechanism is located to the left of the action body. The firing pin is 416 stainless steel with the shaft being 316 stainless steel. The length of the action ensures a smooth cycling process with the twin-bolt lug system and there is a rounded anodized bolt knob for good grip. The action can be supplied with a Timney trigger plus safety, if desired.

John Carr Actions

John Carr owns Revolution Precision, based in Doncaster, and is a recent addition to the list of great British action makers. Custom actions are his primary goal, but he also makes bolt handles, Picatinny rails and muzzle brakes. His actions are stainless steel perfection with barley sugar-twist flutes to the bolt and even titanium nitride-coated bolts. The three action styles – the Clubman, Classic and Premium – cater for different cartridge and action length sizes to accommodate any configuration for the target, stalking or long-range shooting disciplines. As a precision engineer, John can make bespoke parts to 'customize' any action to the client's specific idiosyncrasies and supplies many custom gunsmiths in Britain, including Steve Bowers, Valkyrie Rifles and Steve Kershaw.

The Mayfair is a very English variant on the iconic Mauser 98 action, but made in stainless steel and to benchrest quality tolerances.

Mayfair Engineering

Lovers of traditional Mauser actions for a quintessentially English styling should consider those by Mayfair Engineering. These are designed and manufactured using Computer Aided Design (CAD) to give a traditional appearance, but all the tolerances on the surfaces and concentric faces inside, where it really counts, are held to a minimum. The result is one of the best Mauser-type actions built anywhere in the world.

There are three action sizes: Small for .22-250 to .308 Win sized cartridges; Standard for .275 Rigby to .300 Win Mag; or magnum for large-diameter rounds like the 505 Gibbs. Each action is available in left- or right-hand configuration. To add further versatility, all action lengths are available with a double square bridge, single square bridge or the standard rounded action top, so all scope mounting options are catered for. Actions and magazines are meticulously crafted in the Mayfair workshops and tailored to each individual cartridge to ensure perfect bolt cycling, reliability and cartridge feed. The action is fashioned from a solid block of stress-relieved 8620 nickel-chromium-molybdenum steel on their CNC and EDM machines to the tightest tolerances. Only then is each individual action hand-finished and lapped for the smoothest Mauser 98 action that can be achieved. There is no cut-out to the left action face to improve looks and rigidity. The standard action size weighs just over 3lb.

The bolt is fashioned from a solid billet of 8620 steel, stress-relieved and then CNC machined to duplicate the classic twin-lug, non-rotating extractor and third safety lug bolt profile. These bolts are fully case hardened and can be supplied both with or without a scope cut-out. The non-rotating extractor is machined in three different lengths from very strong and reliable silicon manganese spring steel. The firing pin tip has an outside diameter of 1.7mm for use with high-pressure cartridges. Similar attention is paid to the bottom metal or magazine. It is also machined from a solid block of 070M20 steel and is fine-tuned to each cartridge's size and dimensions. All the angles and shoulder cut-outs are perfectly machined to match the individual cartridge, providing exceptionally reliable feeding.

Trigger guards can be offered on the standard Hunter or Safari profiles. The all bottom metal has a four-degree draft angle around its total profile edge, which really helps to inlet the action accurately into the stock. All floor plates are hinged and opened by a simple push activation. Finally the safety is modified to a three-position, scope-friendly, lever/wing type, a departure from the classic Mauser design that nonetheless is an improvement.

In-house Designs

Many custom rifle makers use the four British action makers mentioned above to produce a custom-made and designed action to their own specification, which allows better lead times than waiting for actions to be sourced from America. It also means you can have a truly unique action for your rifle, perhaps ordering a left-hand bolt, a right-side port with a helical fluted bolt, polished action and faceted action sides. The Impala actions from the Anglo Custom Rifle Company offer very high precision and elegant action designs for all cartridge lengths and orientations. The Valhalla and Thor actions from Valkyrie are beautifully precise and smooth actions that will grace any custom project, offering superb accuracy. Steve Bowers also collaborates closely with British action makers to produce specifically designed actions to his own specification.

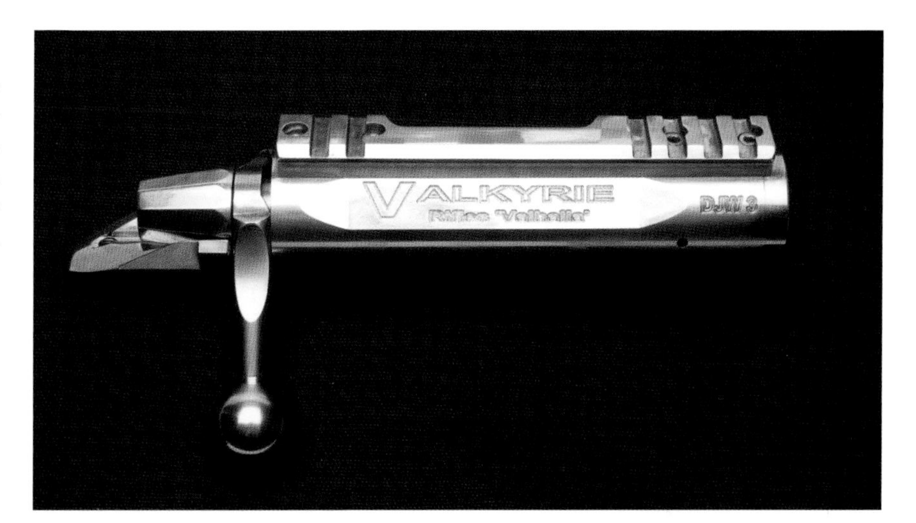

The Valhalla custom action from Valkyrie Rifles is David Wylde's own take on a Carr action. Both the Valhalla and the smaller Thor action are superb.

Impala actions, made for the Anglo Custom Rifle Company, will also adorn any custom project.

Chapter 3
The Barrel

The barrel is the most crucial part of any custom rifle. It is directly in contact with the bullet as it spins down the rifling and is responsible for containing the explosion generated by the powder ignition. All this happens in a split second; if you get any part of it wrong, you won't hit your target. The barrel is not the only contributing factor to accuracy or consistency, but if you economize too far or get the barrel specifications wrong a custom rifle will not shoot, regardless of the action or the stock. That's why barrel, even blanks, are expensive. You should allocate as much of your budget as you can on this component and its subsequent chambering and profiling.

Erosion and corrosion can have a serious effect on a barrel internally, but this is your only option in guiding the bullet from gun to game. If there are any problems a bad barrel will not steer a bullet true and any imperfections will be magnified downrange. It takes a mere fraction of a second from when you think about taking the shot and actually squeezing the trigger to the bullet leaving the barrel.

- Between thinking about shooting and your brain sending this action to your trigger finger's nerve endings takes at least 0.2s.
- The trigger releases the firing pin, which connects with the primer and ignites the powder, taking another 10 thou of a second: 0.21s.
- The powder turns to gases and starts to exert its pressure on the base of the bullet: 0.211s.

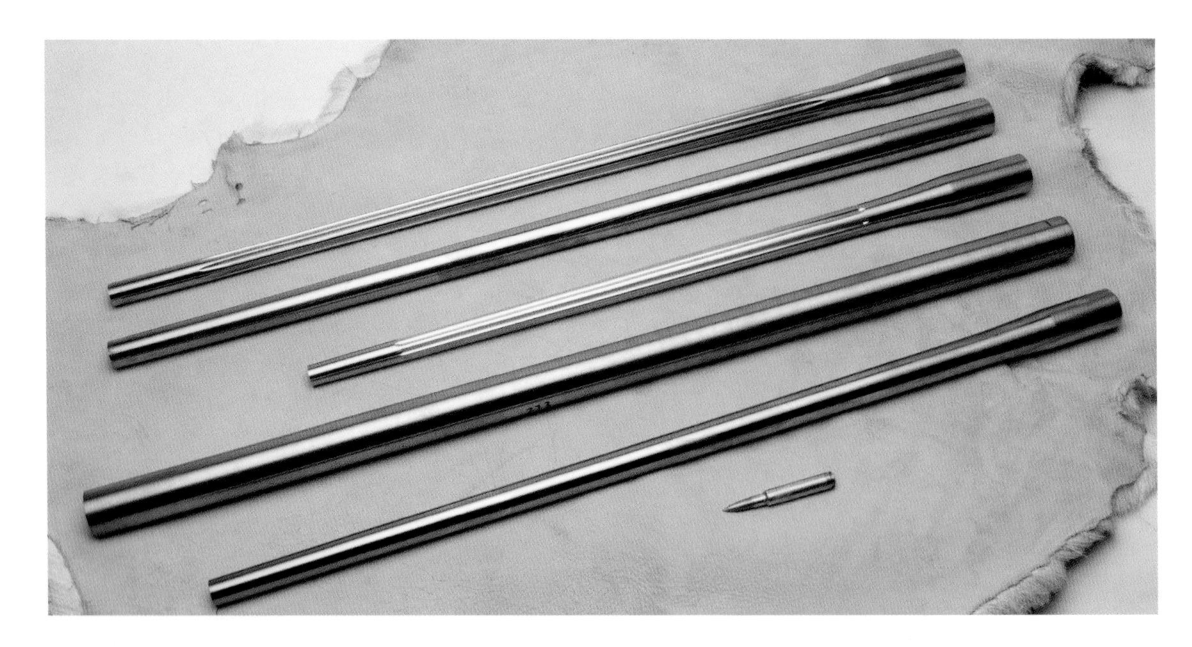

The barrel on a custom rifle project is the key to the whole process. Although it may be expensive, it is essential to choose the right type for your shooting needs.

The whole firing cycle takes place within a split second, so every part of a custom rifle has to be in perfect harmony to achieve the accuracy you want.

- The bullet now moves up the barrel about 4–5in up the barrel and is accelerating: 0.212s.
- On a 20in barrel for a 140 grain 7mm bullet travelling at 2850 fps it leaves the barrel: 0.2124s.
- If the target is 100 yards away it will take a further 0.145s to get there: 0.3574s.
- At 0.425s to 0.445s you will feel the recoil.

At any point in the firing cycle where the bullet is in the barrel it can be adversely affected by poor bedding, barrel movement, sluggish lock time, a bad trigger or inconsistent rifling. That is why every part on a custom rifle needs to harmonize together to provide an identical firing cycle that translates to consistent performance and hence accuracy. When shooting lead bullets using a low-pressure black powder load, the pressure in steel barrels would reach about 25,000 pounds per square inch (psi) and corrosion from the salt-based black powder used to be an issue. Compared with this, modern centrefire cartridges can initiate pressures in excess of 70,000 psi when pressure-proof tested at the proof house. The barrel also has to contain the extremely hot gases caused by the burning powder that erode the bore's surface.

Heat and erosion are further accelerated by the frictional forces of a fast projectile with a copper jacket around a lead core. Therefore it is essential that the barrel steel you use is machined to a smooth internal surface to ease the bullet's passage, is stress-relieved for accuracy and, importantly, has a hardness level for long life of at least 28–34 RC. The steel selected should be of 4140, 4142 or 4150 standard, which is defined by the alloy content of elements such as manganese, silicon, nickel, chromium, molybdenum or vanadium. Manganese increases the strength and hardness of the steel. Silicon raises the temperature for heat treatment and adds resilience. Nickel helps with corrosion resistance, as does the chromium used in stainless steel, and it also increases hardness. Finally molybdenum is used to increase high-temperature tensile strength and hardness.

It is important to use a barrel from a premium supplier but it is the way the barrel is bored, chambered, fluted, lapped or rifled that really makes the difference. It is the same with scope lenses: there are only a few big lens and prism makers supplying the industry worldwide, but what makes a real difference is the way the lenses are coated and polished. Most custom barrels are held to a tolerance of at least 5/10,000ths of an inch and that is the quality you are paying for. The straightness of the bore and the uniformity of the rifling, demonstrated in the rifling lands, depth, width and twist rates, are key factors that really affect accuracy. In a nutshell, you need end-to-end perfection and uniformity. The internal surface of the chamber and the throat

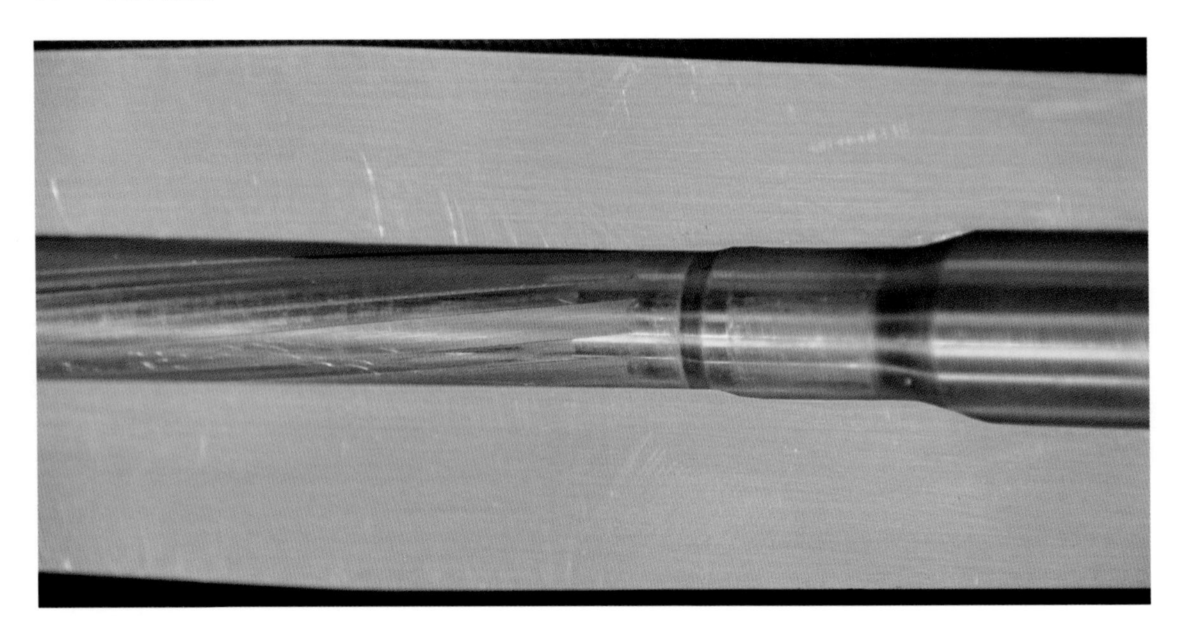

A custom rifle must do more than look good on the outside. Anything else is meaningless if the internals, especially the bore, are not perfect.

area must have no rough tool marks, otherwise fouling will be too great and destroy accuracy. Their direction to the rifling and throat area as the rifling begins is crucial. Perpendicular reamer cuts or marks across the lands will foul and upset perfect bullet performance.

Even when satisfied that the bore is perfectly straight, the next problem is ensuring a perfectly chambered barrel that is concentric to the bore and also concentrically threaded to the action. Unless this is achieved all the work on your expensive custom barrel has been wasted and it will never reach its true potential. If the bullet, the only part of the rifle that actually hits your target, does not start its journey down the barrel in a straight line because of a crooked chamber, then you might as well give up. A poor non-concentric bullet in the bore will become deformed and never centre correctly in the barrel, rotating around a different axis to that of the centre of the bore. When it leaves the barrel it will continue to rotate around a new central point and follow a spiral path rather than spin on a straight axis. If you achieve perfection in all these factors, the sum of all the parts in your rifle will result in the harmonious union on which a fine shooting rifle depends.

BARREL MATERIALS

Chrome Moly 4140, 4150 and 4340 Grades

This is the favourite material used by classic rifle makers for the best traditional custom rifles. It is primarily used on hunting rifles and not target weapons. The main reason that chrome moly is so popular is that it takes a deep rich blue to the surface, the preferred finish to any classic rifle. Chrome moly barrels commonly consist of a chromium molybdenum alloy with vanadium added for strength and long life. Chrome molybdenum, also known as chrome moly vanadium or chrome moly, has a good tensile strength and is cheaper than stainless steel, which makes it suitable for many other manufacturing parts such as gearing. Despite its name it does not have a high chrome content and so is less resistant to rusting than stainless steel. The number used to indicate a grade is usually contains four or five digits: the first number relates to the type of steel, such as carbon or nickel for instance; the second number indicates the alloy

Custom rifle makers like Callum Ferguson have to keep a lot of barrels in stock to cater for their customers' varying requirements for stainless or chrome molybdenum from various makers, and a choice of calibre and rifling twist rates.

added and its percentage; and the third and fourth figures indicate the percentage of carbon content. A figure of 30 or below is a low carbon content, while a higher number shows that more strength has been added to the bar stock. A typical barrel of 4140 steel can have a tensile strength of 160,000–180,000 psi when tempered, whereas standard steel is more like 90,000–100,000. Its 40 per cent carbon content is also supplemented by small amounts of chrome, manganese, molybdenum, phosphorus, sulfur and silicon.

It is often said that stainless steel is better and more accurate, but chrome moly barrels are actually the equal of a good stainless barrel up to match grade. Chrome moly barrels often have a more burnished and smoother interior and respond well to hand lapping, giving a mirror-like finish that is not only precise but creates a good gas seal behind the bullet, helping to avoid excessive copper fouling. Barrel grading will differ from maker to maker. Cheap barrels are cheap for a reason: the bore's interior tolerances may not be totally uniform along their entire length and the bore can be off-centre to the outside diameter of the barrel. This causes stresses in the barrel as the bullet moves down and these inconsistencies will affect the accuracy. The minimum requirement on any barrel destined for a custom rifle should be match grade: the groove diameter should be at least 3/10,000th of an inch uniformity along the entire barrel's length. Also the groove diameter should be within tolerances of 5/10,000th of that of the bullet's diameter. This is why custom barrels cost more than standard barrels, as custom makers hold much tighter tolerances and guarantee their work.

Stainless Steel 416 Grade

Custom rifles destined for target or silhouette shooting, long-range varminting or deer stalking are usually made with stainless steel barrels to provide longer barrel life. This is especially so when using higher-velocity calibres that push small bullets down tiny bores with

The RPA switch barrel rifle I use for testing my wildcat creations. Each barrel is stainless steel and smithed by Steve Bowers.

very hot gases, eroding the rifling more quickly. With bigger calibres there is less advantage to using stainless steel, but the choice is very subjective due to the actual calibre and powder charge behind the bullet and how the rifle is fired, given that repeated rounds without letting the bore cool properly can ruin any barrel. Stainless steel also looks good on a custom rifle, but to many the real advantage is that out in the field stainless steel lives up to its name, being rust, corrosion and erosion resistant, and requires less exterior maintenance. It is foolish, however, to think the interior surface does not need the same cleaning regime as a chrome moly, as it does most definitely.

There is a greater demand for stainless barrels than for chrome moly and the former often come in more grades. A good match barrel will equal the best chrome moly barrels, but for a little extra the select match-grade barrel offers the top or best quality available. As the name suggests, these barrels are individually gauged: those that hold a groove size to within 3/10,000th of an inch with a uniformity of 1/10,000th of an inch along the entire barrel are selected and classified as select match grade. Whether these shoot any better than a match grade barrel is subjective, but there is no harm in getting the best you can. It is always best to start at the top and eliminate any variations from the beginning. If you have a cheaper barrel, all that custom work can be for nothing if the barrel will not shoot.

The 416 stainless steel used for barrels has a higher sulphur and chrome content than the more malleable 410 stainless steel found in frying pans and pots, and has superior qualities for machining and corrosion resistance. It is more expensive than chrome moly, but many will deem this well worth it.

Carbon Fibre

Barrels employing carbon fibre were made famous by Christensen Arms. For rifles requiring both a lightweight feel and a heavier barrel profile, a slender steel barrel is wrapped in carbon fibre to stiffen the overall barrel without increasing its weight. The barrel starts as a button-rifled select match-grade blank that

Paddy Dane Benchmark carbon fibre-wrapped barrel fitted to one of his custom rifles for a lightweight hunter.

is then turned down to a profile that suits the intended cartridge size. The carbon fibre wrap is then applied to restore the original barrel profile. The advantages of this technique are that it saves weight and improves the structural integrity and thermal properties. Very thin and slender steel barrels used extensively for mountain can act like a whip as it heats up and accuracy is lost. A slender steel barrel embedded in carbon fibre, however, remains rigid and heats up less quickly. A portion of barrel is usually left at both ends for chambering into the action and fitting a muzzle brake or sound moderator. The difference in weight between a standard steel barrel and a carbon fibre barrel of the same outside diameter or profile can be as much as 1.5lb.

Carbon fibre barrels offered by the Anglo Custom Rifle Company are custom made to order in the USA by K.K. Jense at Jense Precision. Select match grade stainless steel barrels are wrapped with carbon fibre to the correct profile and length. Paddy Dane from Dane and Co. Ltd also offers carbon fibre barrels on his custom rifles. The benchmark carbon fibre barrels with MTU contour makes a lightweight rifle in any calibre. Paddy also stocks of a promising new design of carbon fibre barrels by Proof Research.

CUT OR BUTTON RIFLING

Cut Rifling

This is considered to be the best form of barrel making. It is the traditional method in which the rifling is made by many passes from a cutting tool to form the grooves down the bore. It is commonly called a single-point broach-cutting process. A cutter tool that resembles a hook is placed inside a steel cylindrical retainer (cutter box) that is made to fit and travel down the reamed barrel blank. This in turn is pulled through the barrel by a steel tube that administers cutter fluid, thus allowing the broach cutter to cut a single rifling groove at a time. The rifling twist rate is governed by rotating the cutter box on each pass. Depending on the number of grooves cut, each groove is cut in sequence until the correct groove depth is reached.

It takes much longer to cut a rifled barrel than a button- rifled one, which is why cut rifled barrels are more expensive. There is also the benefit that the tools used to cut these barrels all work in parallel along the barrel length. This means that most of the tool marks run with the rifling or grooves and not across them, which helps maintain good accuracy, less fouling and keeps the bullet jacket undamaged. Cut rifling also has the very real advantage that you are tak-

Cut rifled barrels are always perceived to be the most accurate type. Steve Bower's machining cutter boxes for a client's cut rifling business ensure true precision.

ing only a small amount of metal away from the bore at each pass so you are not stressing the barrel, which can cause inaccuracies. This also means that fluting the barrel will not cause any alterations to the barrel profile or harmonics, and no stress-relieving procedure is needed. The greater internal concentricity means that the bore axis is very precise. A cut rifled barrel also has a precise and accurate twist rate. If a 1 in 10in twist is specified then that is what you get; if the button slips you might finish up with buttoned barrel with a 1 in 10.3in twist. A cut barrel also gives the option of a gain twist barrel, in which the rifling twist can start slowly for lower pressure and then increase in rate to a fast twist. This provides final stabilization of a heavier projectile than is stabilized in a barrel that has only a slow twist rate to start with. Krieger, Bartlein and the new GB cut-rifled barrels, for example, can offer extreme accuracy for any custom rifle project.

Button Rifling

Button rifling is probably the simplest and quickest method of rifling a bore. It is a surface forming method that does not involve removing metal when the barrel is rifled. It uses a swaging button made of carbide to impress the rifling into the internal bore of the barrel. This is a cold forming process as the lands are cut into the carbide button and then pulled through the barrel's bore with a hydraulic press. thus impressing the rifling into the barrel. Different buttons are used to vary the rifling twist rate, due to the angle of the imprinted rifling form, so that switching twist rates is cheap and quick. The button is made several thousandths of an inch larger than the finished bore size as the elasticity of the barrel's steel ensures that it springs back after the button has passed through the bore.

Maintaining a true rifling twist can be tricky as the button can slip or drag, resulting in a twist

This button rifling machine at Sassen Engineering produces Border barrels, which are of exceptional quality.

rate more or less than stated. This is fine if it is consistent all down the bore, but may be a problem if it differs along the barrel's length. There is also a propensity, especially if the barrel is not totally uniform in its internal crystal structure and has small pockets of impurities, for the rifling not to be engraved uniformly on both sides when the button is passed through the bore, resulting in deeper and shallower grooves down the barrel. This, however, can also be a problem with cut rifle barrels.

Although these differences are microscopic, if the overall concentricity of the bore is flawed it can still cause a bullet and the expanding gases to destabilize and erode or foul a barrel. The depth of grooves can also be influenced by the diameter of the barrel. If it is button rifled after it is profiled, then the receiver end will have a large diameter of steel compared to the muzzle end. Now if the barrel is button rifled as it passes down the barrel at the receiver or chamber end, it will engrave deeper grooves as

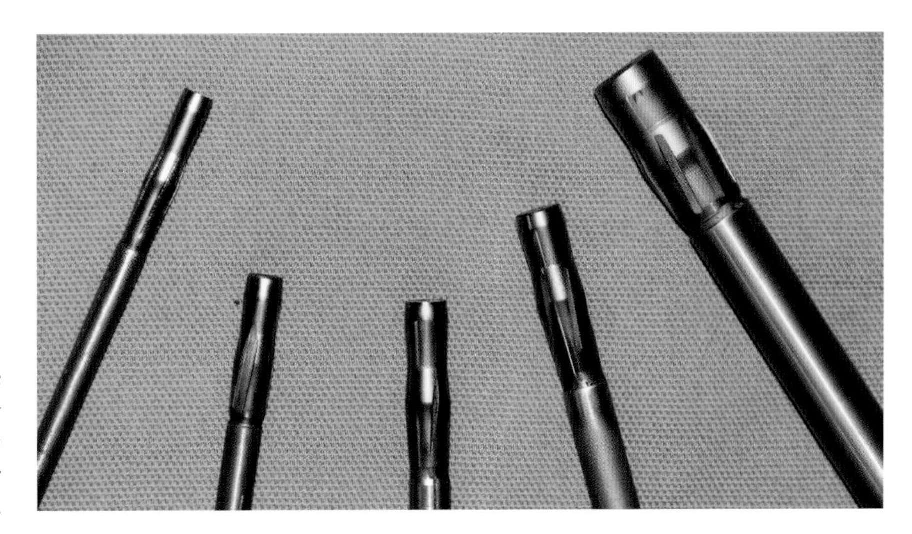

Different calibre buttons are used by Sassen Engineering when rifling their Border barrels.

there is more metal and thus more resistance to the passing button compared to the more slender portions of the barrel towards the muzzle, which can produce a shallower groove due to the metal being elastic enough to expand as the button passes. That is why button-rifled barrels are best rifled as a solid blank and then profiled, but stress relieving is best performed before profiling. The advantages of button rifling are that it is cheaper and faster, and that it is easier to make odd sizes, rifling styles, shapes and configurations. It is also more suited to unusual rifling twist rates, including extra fast twists, which are difficult to achieve with cut rifling in small calibres, and speciality rifling such as polygonal rifling twists. Shallow grooves are easily formed for subsonic use where extreme fouling from copper jackets can be a problem. It is also easier to have more lands or grooves for specialized rifling contours. A button-rifled barrel has surfaces that are smooth, regular and almost work-hardened, and so shoots well from the start, although hand lapping would still improve its accuracy.

A button usually work-hardens the bore by burnishing the inner surfaces so that they last longer time and foul less. Chrome moly barrels are easier to button as stainless steel tends to gall more easily when the steel rides up into the rifling grooves and clogs. I know people can get snobby about cut rifling barrels being the best, but from a personal point of view I have owned more good button-rifled barrels than cut-rifled ones.

Sassen Engineering, based in Birmingham, makes a complete range of button-rifled barrels that are very consistent, straight and beautifully made, and shoot very small group sizes. A special honing machine is used prior to button rifling, which ensures the internal bore surface is as uniform and perfect as possible. When the button is passed through the result is a perfectly rifled bore with none of the normal associated problems. All the barrels are correctly stress relieved in a purpose-built oven.

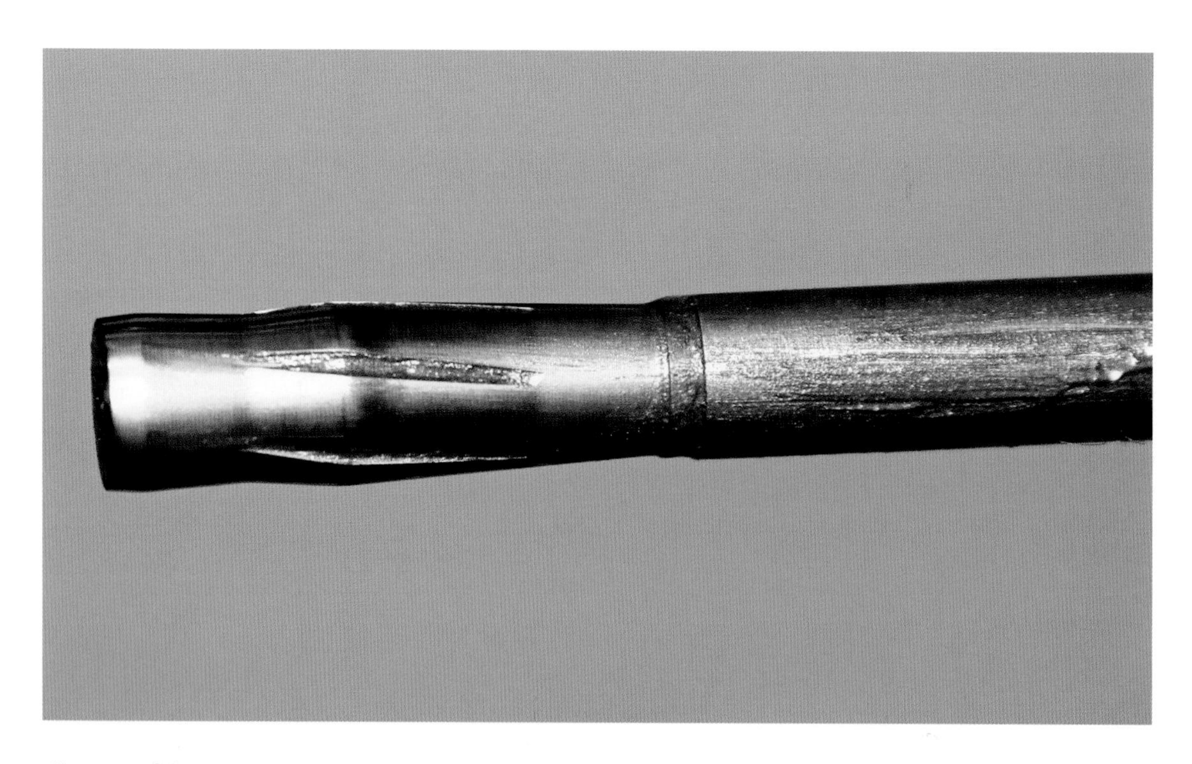

Close-up of the Sassen Engineering button, showing the raised rifling profile used to imprint the rifling into the bore.

Lewis Miller, the custom barrel maker at Sassen Engineering, using their advanced bore-honing machine to produce a barrel of superb uniformity.

HAMMER FORGING

Hammer forging uses a method in which the barrel is bored to size and then hammered around a tungsten carbide rod with the exact reverse image of the rifling in it. This forms the rifling down the barrel as the carbide mandrel is passed down the barrel blank. This is actually shorter than the finished length as the barrel blank is hammered repeatedly around the mandrel rod as it is passed through the forge machine. A uniform rifling imprint is made inside the bore by sets of radially placed hammers around the mandrel. It is an expensive process but has the advantage that it can produce barrels very quickly. It was first used for machine gun barrels due to the high rate of attrition caused by the high cyclic rates of the weapons. Problems can result from the heavy stresses imparted into the steel barrel in the hammering process. One benefit, though, is the internal surfaces are work hardened. Hunting rifles by Steyr use hammer-forged barrels that share the characteristic spiral external contours common to Steyr rifles.

NUMBER OF GROOVES

The grooves and rifling land profile are exceptionally important to any custom barrel project as this is the part of the bore that contacts with the bullet's jacket, so it has to be correct. The number, size, depth and profile all play a part in the stabilization of the bullet's passage down the bore. Typical rifling will have five to six grooves, which means between 25 and 35 per cent coverage of the actual bore surface. Finding a working combination of number, size and depth is an art. The results will be specific to calibres and bullet weights. If the rifling land is too slim and the groove is wide, the rifling becomes susceptible to erosion from the hot

Typical Grooves per Calibre

Calibre in inches (mm where indicated)	Grooves typical	Grooves optional
.14 (.144)	4	3
.17 (.172)	5	3 or 4
.19 (.198)	6	3 or polygonal
.20 (.204)	3	4 and polygonal
.22 (.224)	5	3, 4, 6, 7 or polygonal
.22 rimfire (.222)	6	3, 5, 9 or polygonal
.243 (6mm)	5 or 6	3, 4 or polygonal
.25 (.257)	5	3, 4, 6 or polygonal
.264 (6.5mm)	5	3, 4, 6 or polygonal
.270 (.277)	5	3, 6 or polygonal
.284 (7mm)	5	3, 6 or polygonal
.308	5	2, 3, 4, 6 or polygonal
.303 (.311)	5	3
.323 (8mm)	5	3
.338	5	3, 6 or polygonal
.35 (.358)	5 or 6	4
.366 (9.3mm)	5	polygonal
.375	5 or 6	3 or polygonal
.411 (.410)	8	6
.416	6	3 or 8
.458	8	3 or 6
.510	6	3, 8 or polygonal

expanding powder gases and will wear quickly. At the same time, though, it will not contact the bullet as much as a wide land and so will not foul as greatly. Wider rifling lands means that fewer will fit in the space: three or four lands are less affected by flame erosion than thin ones and are therefore popular on smaller fast .22 and .20 calibre rifles.

I find that three-groove rifling on my small-calibre, fast-velocity .17 and .20 calibre rifles or wildcats is accurate with uniform twist rates along the bore. They also tend to foul less with speedy calibres like the .17 PPC, .20 Satan and .22-284, and pressure levels seem lower than with more rifling lands. This is when a custom rifle can be recognized for more than its external beauty. If internally the twist rate and rifling dimensions are perfect then you have a true custom rifle – it's just you cannot see the effort made inside the barrel.

Subsonic bullets travelling below the speed of sound require very different rifling dimensions. They not only require a faster than normal twist rate to stabilize a bullet around its axis, but very slim, shallow and numerous rifling lands are beneficial. This is primarily due to the slow movement of the copper-jacketed bullet down the barrel, which deposits copper on the lands and into the grooves more readily than when shot supersonically. Attempts have been made to use lubricants such as tallow, wax, moly and even Vaseline to eliminate copper fouling, but the best solution I have seen is that adopted in SSK barrels. J. D. Jones of SSK Industries revolutionized subsonic cartridge design with a range that went from the

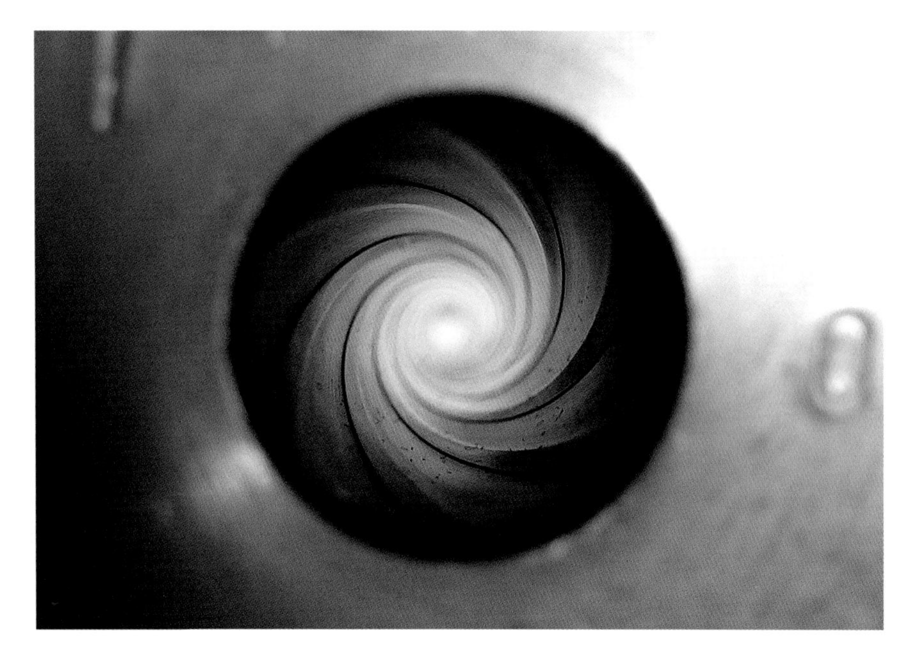

In order to stabilize a bullet for accuracy and reduce fouling and bullet jacket torque, the number of grooves and rifling twist rate are crucial.

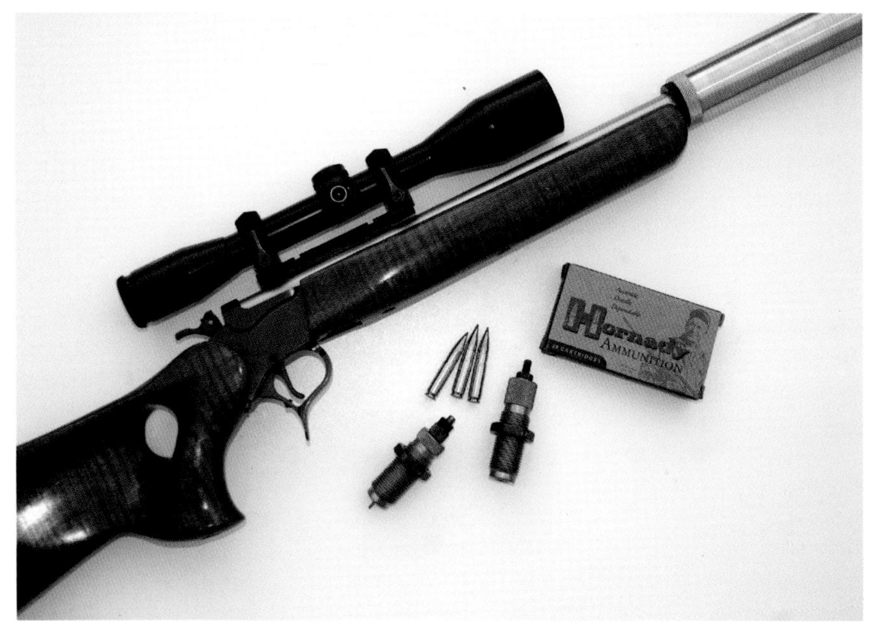

The 300 Whisper by J. D. Jones launches heavy .30 cal bullets at subsonic velocities. This needs a very fast twist and specially grooved rifled barrel to stabilize the bullet and reduce fouling.

Micro Whisper, which shot a 180 grain bullet from a .30 Mauser pistol case, through to his famous 300 Whisper round and the 500 Whisper. He significantly reduced copper fouling problems by using barrels with fast twists, but with a larger bore diameter and eight shallow grooves for the .30 calibres instead of the four or five broad, deeper rifling lands in conventional designs.

Tight bores are also popular with some as they produce more pressure, resulting in a higher velocity from a shorter barrel. Tight bores on a .308 Win would be 0.298in bore diameter and 0.307in groove diameter, instead of the standard 0.308in groove diameter. Mike Norris uses a tight-bore system on his Ratal rifle, which allows a shorter barrel but still maintains excellent velocity and accuracy.

Polygonal rifling profiles are a variation on conventional rifling design. Steve Bowers shoots one such barrel in his fully silenced .308 Win subsonic/supersonic rifle.

POLYGONAL RIFLING

This type of rifling is unusual as it does not have the traditional hard-faced edges to the rifling lands. Instead the rifling is smoothed out to a multi-faceted, hexagonal or octagonal form that looks like a spiral flute. These subtle grooves give less torsional grip on the bullet's jacket and are supposed to aid in a better gas seal between bullet jacket and barrel.

I have used several polygonal barrels from Walther and Pac-Nor that certainly exhibit fine accuracy, but whether they are better than normal rifling remains to be seen. One area where they have proven to be advantageous is in subsonic loads, where less barrel fouling is evident. The purported greater muzzle velocities and increased accuracy remain very subjective, but there is reason to turn down a custom polygonal barrel if you want something different.

RATCHET RIFLING

This is where the traditional rifling land has been redesigned to resemble a slow gradient to one edge of the rifling land instead of the more traditional sharp angle. It has this leading edge on the side of the twist rotation to smooth out the bullet's engagement as it enters the rifling lands. In fact the leading edge is almost tripled in length to allow a smooth lead up to the top of the rifling land. This minimizes bullet deformation yet maintains the correct obturation of the bullet into the lands and aligns it smoothly and consistently so that it is set up perfectly around the bore axis. Shilen barrel makers make ratchet rifling barrels.

A similar type of rifling or with the same ethos is that promoted by Broughton Barrels, whose 5C (canted land) barrels are designed to produce more velocity than conventional rifled barrels yet foul less. If you compare barrels of the same length with the rifling groove number the 5C profiled barrels can achieve higher velocities, often up to 100 fps more, and that can make a difference at long range with longer barrels. The Broughton 5C land profile is similar to the 5R rifling designed by Boots Obermeyer, but Broughton uses a precise button and guide system to profile the rifling into a smooth bore surface. The actual angle of the rifling takes up more of the surface of the land, whereas the Shilen ratchet system has a more defined contour only to one edge.

Stress relieving is crucial on button-rifled barrels to ensure that when heat builds up after the shot it does not cause the barrel to shift, destroying accuracy.

It looks a bit like a raised polygonal barrel and, because it is buttoned and easier to use than cut rifling, can be offered in many twist rates to accommodate light or heavy bullet weights, further enhancing its usefulness.

STRESS RELIEVING

A very important aspect of the custom rifle is the need to stress relieve the barrel after all this machining work to rifle the barrel. Stress relieving simply involves placing a barrel into a special oven and heating it with all the air taken out to restore the barrel steel's crystalline structure or relax it back to a uniform consistent state. This procedure is essential to ensure top accuracy as a stressed barrel can alter the point of impact of your shots as they heat up and warp due to weather or heat build-up. If a steel barrel becomes stressed as it is machined or profiled, any resulting movement will become evident in the changing internal dimensions of the bore and grooves, which are directly related to the lower resistance holding the inside in place as weight comes off the outside.

HAND LAPPING

This is a very important part of the custom barrel process as it finishes off the interior of the barrel's bore, removing tool marks and imperfections, and will certainly contribute to excellent accuracy, long life and less fouling. There are many ways to lap the barrel, but the most effective is always by hand as you get a feel for the bore's surfaces through the lapping rod. Usually an exact cast of the bore diameter is moulded from molten lead and attached to a lapping rod. The barrel is then set up in a vice and molten lead applied to the lapping rod, before passing the lapping rod up and down the barrel with uniform and even strokes. This is repeated until all the machining marks are removed and the bore is smooth and bright.

This process finalizes the internal dimensions and polishes out any rough tooling marks. It is impossible to remove all the tooling marks but hand lapping certainly helps reduce fouling in the barrel that contributes to poor accuracy. A consistent and smooth interior bore surface also eases the bullet's passage and makes it more likely to remain stable and spinning concentrically about its core axis. Sassen Engineering uses both hand lapping and computerized lapping machines to finish their barrels.

As a final process Sassen Engineering's Lewis Miller always hand laps the Border barrels with a lead lapping tool to ensure an ultra-fine finish.

PROFILE AND LENGTH FOR OPTIMAL PERFORMANCE AND APPEARANCE

Profiling the barrel is another crucial factor in a custom rifle's make-up. If you make a mistake the barrel may be too heavy to be carried all day on mountainous terrain, or the barrel might heat up because it is too light and accuracy is lost. In practice you need to find a compromise. If a barrel is long it usually needs to be heavy profiled to maintain rigidity and reduce the tendency to flex and heat up too quickly. If it is short, again a heavy profile is beneficial as it makes the barrel more rigid, suppressing harmonics and making it less susceptible to heat transfer, which upsets accuracy. There has to be a compromise, however, as too short a barrel will not allow the whole powder charge to burn and cause a muzzle flash and lower velocities, while too long a barrel will destroy handling.

You have to determine the true end purpose of your new custom rifle before you can make a good judgement on the barrel profile needed. If you intend to use your rifle solely as an infrequent hunting tool, then only one shot should be all you need and a lighter-weight barrel profile will suit you. If you intend to shoot long range with high-velocity cartridges, often shot from a static or prone position, then you should be looking at a long and heavy-profiled barrel that can be shot repeatedly without heating up in order to maintain the point of impact or prevent groups opening up.

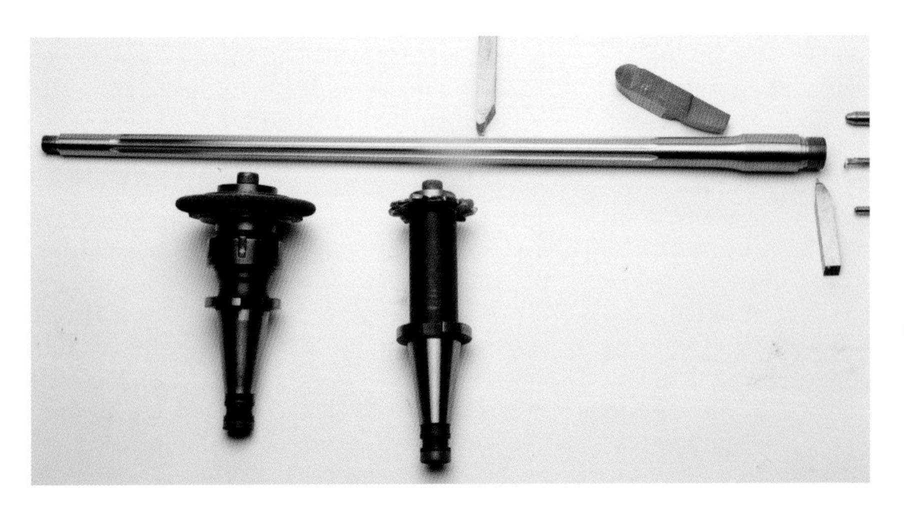

Barrels have differing profiles for varying shooting disciplines. Steve Bowers precisely contours the barrel to the customer's own specification.

Most modern custom gun makers will order the barrel blank with the correct length, rifling twist and calibre, and also specify the profile required. Few gunsmiths actually reprofile a barrel blank these days unless the customer requires a special barrel profile for a specific purpose, such as weight restrictions or fitting a silencer. Plenty of metal is removed when profiling a barrel from a blank, and if any stresses are present in the blank it can then bend or distort as the metal is removed. If you profile too excessively the groove diameter can increase as you remove the holding force of extra metal around the bore. If your barrel has a lightweight barrel profile, then you run the risk that the bore can be several thousandths of an inch larger at the thinner parts of the barrel, especially at the muzzle. This is even more likely if it is threaded for a sound moderator.

Harmonics are the key to accuracy and consistency. As a bullet passes down a barrel, the frictional forces between the bullet's copper jacket and the barrel's steel rifling sets up incredible torsional and rotational stresses. Each rifle behaves differently, depending on the bullets used, calibre, length, weight, style and velocity. When a bullet is fired I think of it as a car starting up without any oil in the engine, bare metal to metal. This is where the wear takes place. More importantly, however, the bullet's passage down the barrel for that particular load is unique. It is like when you want a car engine to hit a sweet spot where all the parts are in unison with each other, or harmonically tuned, and at some point the ride smoothes out and the steering wheel stops wobbling. The same is true with the barrel. Find that sweet spot and your accuracy will be perfect or as good as it gets.

We call it barrel timing. This is because the barrel's resonances move as the bullet travels down the barrel. In fact the barrel bulges ever so slightly in a ripple effect as this sets up a wave form. Every barrel and bullet combination has a unique wave form. The idea is to get a bullet to leave the barrel at exactly the same time and at the same point of the barrel wave. This is because at the top or bottom of the curve the barrel will be slightly shifted out of true from the original bore axis. Barrel length, weight and profile will drastically affect the way the barrel harmonizes and hence the wave form, but for practical purposes the need to offer a wide range of styles is essential for differing shooting disciplines.

Barrel timing is when all the rifle's components are working together to maximize accuracy. Once you have an accurate reload, the QuickLOAD ballistics program can be used to predict all other reloads to that barrel timing in order to ensure the highest accuracy with differing load components.

Barrel Contours (Shilen)

Number	Type	A	B	C	D	E	F	G	Weight lb
1	Featherweight	1.20	0.690	0.550	2.5	7	20	26	2.5
2	Lightweight	1.20	0.760	0.575	2.5	7	24	26	2.75
3	Sporter	1.20	0.780	0.625	2.5	7	24	26	3.0
4	Magnum Sporter	1.20	0.835	0.650	2.5	7	24	28	3.75
5	Lightweight Varmint	1.20	0.900	0.700	3.0	7	26	28	4.0
5.5	Medium-weight Varmint	1.20	0.950	0.750	3.0	7	26	28	4.5
6	Lightweight Target	1.20		0.750	3.0		26	28	5.0
7	Bull Barrel	1.20		0.900	4.0		26	28	6.0
8	Heavy Bull Barrel	1.20		1.000	4.0		26	28	7.0
9	Rimfire Target	1.125		0.825	4.0		26	28	5.5
10	Straight Cylinder	1.20 to 1.350							

Barrel profile picture that relates to profile dimensions in the table text above and the images above and below.

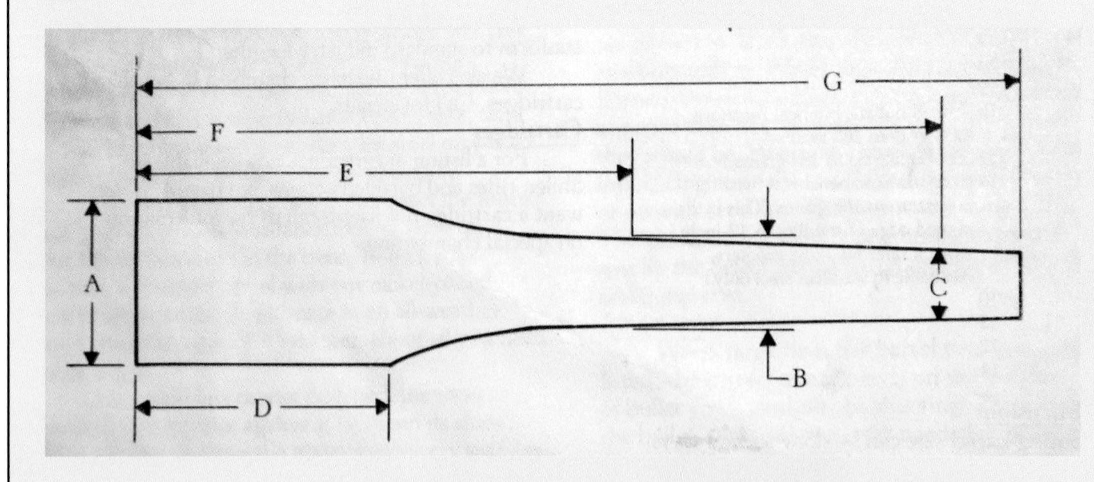

All the outside or profile of a barrel has a marked effect on the overall accuracy potential. The contour of the barrel certainly benefits from a profile that allows the barrel to remain stiff and rigid and less prone to vibrations. Manufacturers have their own in-house profiles and many custom rifle smiths design their own profiles and contour themselves so this is only a guide.

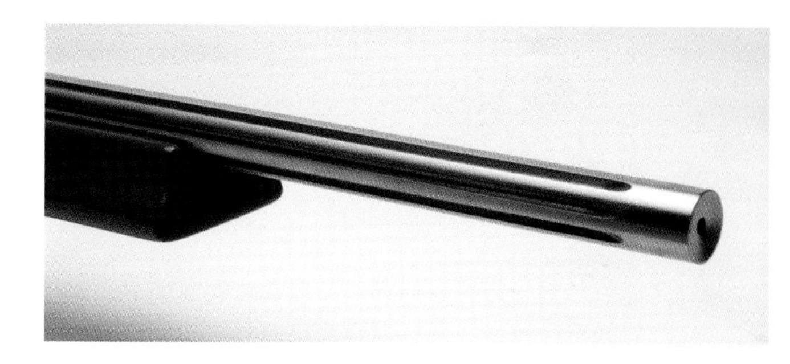

Straight flutes are the most common. They help to reduce weight and give that custom look.

FLUTING

This has become very popular among custom rifle builders for two main reasons: firstly, it looks very good and customers love it, but also, more scientifically, it is possible to have a long barrel for the same weight as a shorter one, while gaining the advantages of a longer barrel for extra velocity. Up to 18oz of metal can be removed in a fluting process, making a number five profiled barrel feel like a number three. (More typically, however, the reduction would be from five to four.) There is also the advantage that the surface area is increased because the flutes produce scallops in the barrel's surface. This helps to cool the barrel and so more shots can be fired before heat build-up ruins accuracy. Much of that is conjecture, but fluting does increase the surface area of the barrel and so is supposed to dissipate heat more quickly, but in truth it depends on the rate of fire. The appearance, though, is improved by the contrast between the colour of the flutes and the main barrel contour, which accentuates the profiling.

Spiral or helical flutes, as on this Valkyrie custom .308 with Cerakote finished flutes, look superb against the stainless steel barrel.

LENGTH

A barrel's length is often dependent on the size of cartridge used and your aspirations for getting the best velocity you can. A long barrel, though, is often ungainly, heavy and impractical for field use when fitted with a sound moderator. Length and weight have a vital role to play in the overall feel and accuracy of a custom rifle. A heavy barrel will have

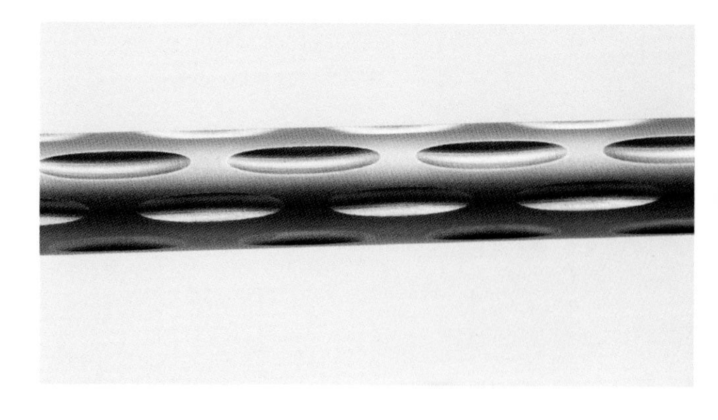

The interrupted flute patterns on this Dolphin Gun Company custom barrel give a very classy appearance.

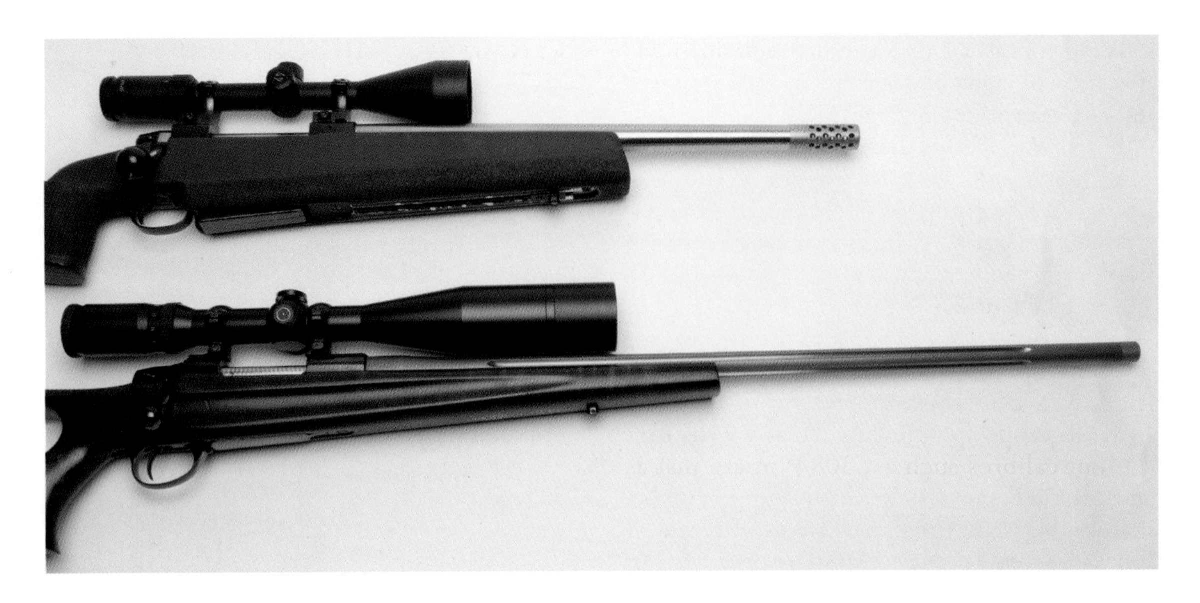

The length of a barrel is largely dictated by the cartridges used and velocity required.

Barrel Length	**Cartridge** .17 Hornet 20gr V-Max 10.5gr RL7	.20 Tactical 32gr V-Max 24.0gr of Vit N130	.223 50gr V-Max 23.0gr of RL10X	.243 75gr HP 46.0gr of H4350	.270 130gr Inter-Bond 58.0gr of RL19	7mm Mauser 140gr Ballistic Tip 43.0gr of Varget	.308 150gr SST 44.0gr of H4895	300 Win Mag 180gr Sierra Game King 68.0gr of RL17
28in	3623 fps	3954	3491	3403	3195	2858	2907	3109
27in	3598	3926	3468	3373	3168	2836	2887	3087
26in	3572	3896	3443	3343	3140	2814	2866	3061
25in	3544	3865	3418	3311	3110	2790	2843	3034
24in	3521	3833	3398	3287	3087	2764	2821	2998
23in	3484	3798	3362	3240	3045	2738	2794	2976
22in	3452	3761	3331	3202	3009	2709	2767	2945
21in	3418	3720	3298	3161	2972	2679	2739	2911
20in	3381	3679	3263	3118	2931	2646	2708	2875
19in	3342	3632	3226	3072	2888	2612	2675	2836
18in	3300	3587	3178	3022	2842	2574	2640	2794
17in	3255	3533	3141	2968	2792	2534	2602	2748
16in	3207	3476	3094	2910	2737	2490	2561	2698

a greater mass and thus can dampen down vibrations on firing. This in turn adds stiffness to stop barrel movement. A thin-profiled, lightweight barrel will act like a whip on firing. It may be lovely to carry in the field, but prone to heating up rapidly and inaccurate after three shots. Choosing a barrel length should start with your choice of bullet weight and velocity from the cartridge you wish to shoot, taking into account any legal requirements for deer species with regards to velocity, energy and bullet weight.

Most people play it safe with a 24in barrel, but calibres such as .308 Win are just as accurate and efficient with an 18in barrel, as fitting a sound moderator adds little extra length or weight. On the flip side, small-calibre varmint bullets or long-range rifles need a longer barrel to burn enough powder to generate useful velocity to compete or hit a target downrange. It is a balancing act. Often a barrel that is too long for its diameter will be flexible and whippy rather than rigid, so accuracy will not be optimal. A short, fat barrel, though, is rigid and has more potential for good accuracy where long range and velocity are not major concerns.

Here is a rough guide to some more common calibres. The QuickLOAD ballistics program can determine velocity increase and decrease for differing loads and calibres.

CHOOSING THE CORRECT CHAMBERING REAMERS

A reamer is the cutting tool used to chamber your new custom barrel. The barrel's shooting ability and its final accuracy and function will be seriously affected by the way the reamer is made, using the best high-speed cutting steel and perfectly ground to size. For longer use or repeated chambering, or for cutting tough barrel steel such as Lothar Walther, a carbide reamer is certainly worth ordering.

Solid, Rotating Pilots, No Throat and Throating Reamers

Solid

Simple reamers use a solid pilot in front of the cutting veins. This is calibre specific: in the .30-06 cartridge, for example, it will be between 0.2971in and 0.3000in diameter so that it fits into the bore and not the groove diameter. The only problem with a solid pilot is that the diameter is dictated by the manufacturer and may be tight or loose in the barrel you are using. Another problem is that, as the reamer is cutting, the solid pilot spins around the bore axis on the rifling and can cause radial distortion or marking.

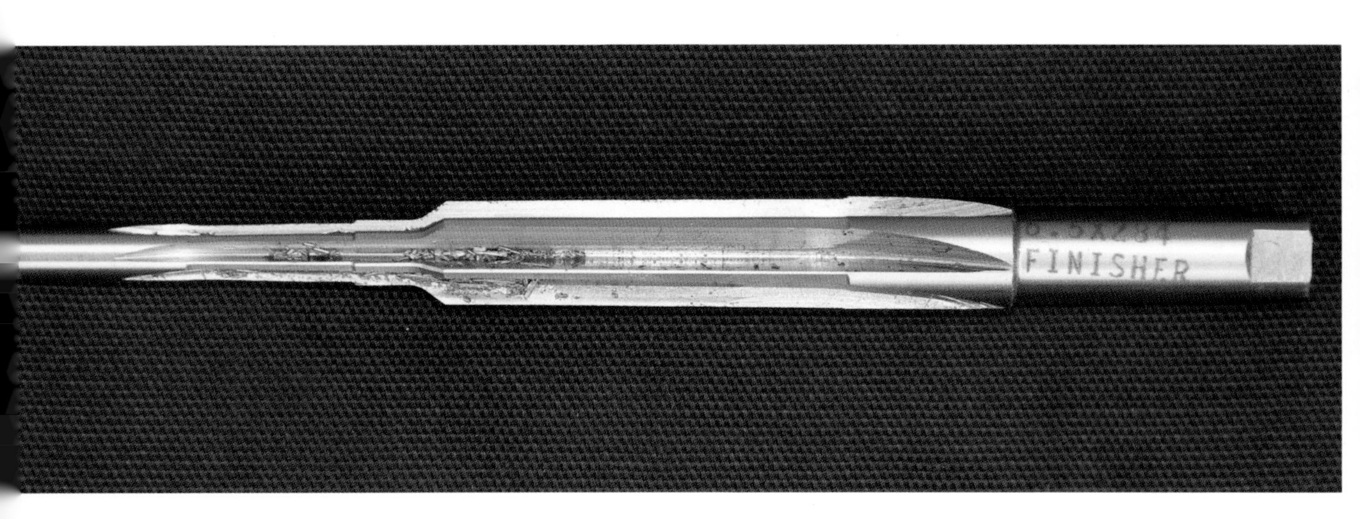

Solid reamers have the disadvantage that the part that enters the barrel's bore does not float freely. This can cause rifling wear or unwanted marks.

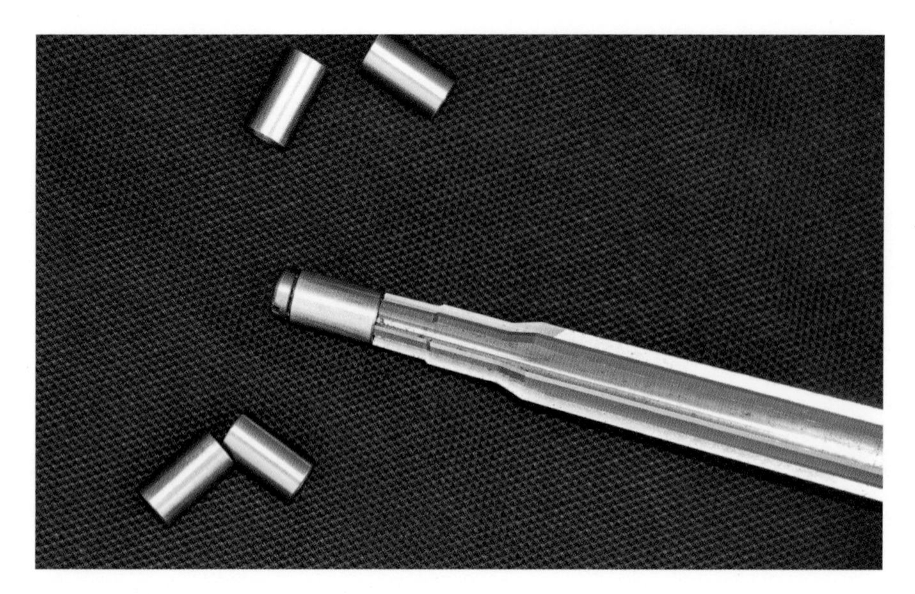

Rotating bushes and replaceable bushes are far better for chamber reaming a barrel as they allow you to match the reamer diameter exactly to that of the bore.

Rotating Pilots

I use these for my wildcat cartridge custom rifle projects. The tip of the reamer has a removable rotating portion that allows the pilots to be added or replaced to match the bore's diameter exactly. In this way the reamer sits perfectly square to the bore and a chamber is cut parallel to the bore's axis. You can also make your own custom-ground pilots to give the custom riflesmith a greater degree of flexibility when setting up a new barrel to be chambered for a precision-cut chamber.

Pre-rougher Reamer

Next you need to choose whether you want a pre-cut rougher reamer as well as a finisher reamer. Pre-rougher reamers are designed to remove most of the metal from the barrel in the first stages of forming the chamber. They typically have a cylindrical shank with large open flutes or cutters for removing metal quickly.

Roughing Reamer

This reamer cuts a chamber to a size just below the finished chamber specifications. It is typically made with dimensions 0.10mm less than the finished chamber. Some prefer to omit this stage, using a pre-rougher or drill to remove the majority of the steel and then running through the finisher reamer as the last step.

Finisher Reamer

This reamer typically has a cylindrical shank and square end for use with a T-bar handle for cutting and guiding. It is used to cut the chamber for a rifle cartridge with rim and forcing cone. The dimensions of the complete reamer are 0.05mm (0.002in) over the minimum standard as set by the Commission internationale permanente pour l'épreuve des armes à feu portatives (CIP, Permanent International Commission for the Proof of Small Arms), which allows a certain degree of regrinding if necessary. It takes a steady hand, proper alignment and lubricated reamers to cut a perfectly concentric chamber, so rushing this stage of a custom rifle project can spoil the barrel. If the reamer binds or digs into the barrel steel, a perfectly good barrel will have to be started again.

Free-bore or Throating Reamer

This tool is used for cutting the forcing cone of chambers of rifle cartridges with dimensions that are exactly as defined by the CIP for every calibre, with a standard angle for the bullet type. Custom gunsmiths order a separate

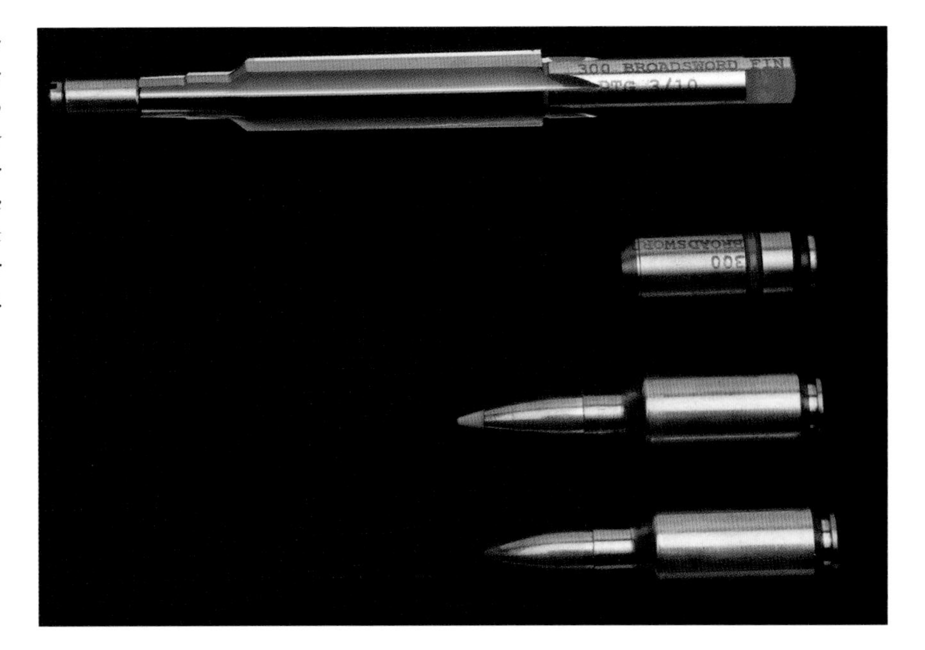

Because Steve Bowers pre-bores the barrel's chambers prior to reaming, I usually order only a finisher reamer and head space gauge, and do not need to use a rougher reamer.

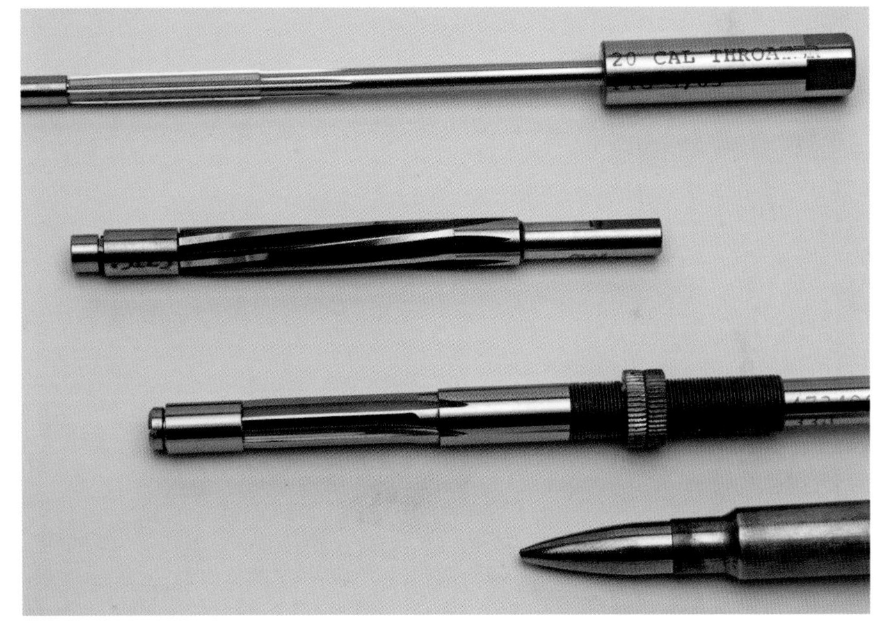

By using a separate throating reamer it can be matched precisely to the bullet length and seating depth required for that cartridge.

throating reamer as this allows the rifle's chamber to be set up to exactly the correct depth to accommodate the bullets specified by the client. The finisher reamer therefore has no throat cutting and this separate throating reamer is then entered into the new chamber to cut the final throat to match the bullets. The throating reamer usually has six teeth for small calibres and ten teeth for larger calibres to ensure the reamer cuts smoothly and leaves minimal tool marks in this vital area of the custom barrel.

Chamber Gauges

These gauges look like small cartridges and are there to mimic the cartridge you want to use in your new chamber. They are made as a pair: go and no go gauges. These are inserted into the

Chamber gauges, such as this 7.92 × 33mm Kurz and .300 Broadsword, ensure that the correct headspace is maintained for safe use.

rifle's chamber and the bolt closed. A go gauge means you have the correct headspace for your cartridge and a no gauge means the bolt will not close as the headspace is wrong. They are made from hardened steel and ground to the exact dimensions of the custom reamer print.

Neck Length and Diameter

Another important aspect of reamer design is choosing the correct neck length and neck diameter for the rifle's chamber. Neck length is a standard dimension when ordered as a normal reamer, but a slightly shorter neck length can be ordered so that, when preparing reloaded cartridges, these can be cut to the new neck length's precise measurements and not to a standard measurement.

Altering a neck length becomes important and beneficial when using non-standard or wildcat rounds. Here you can specify a longer neck to accommodate longer, heavier bullets for the calibre shot as this improves support to the bullet's bearing surface, holding it securely and concentrically. It also allows a greater flexibility to seating depth as a heavier bullet will no longer encroach into the cartridge case, allowing more powder to be added without a compressed load. This is usually also matched to the throat specified on the reamer, so the bullet can be seated closely to the rifling. With a longer neck and a bullet that is no longer seated below the internal neck/shoulder junction, the powder combustion has a better burn characteristic. The gases can flow more evenly around the case sides and walls, and then exert their force squarely onto the base of the bullet. This is not possible when a bullet sticks down into the powder column since the gases can push against the sides as well as the base of the bullet. You also get more burn inside the neck of the case and not in the crucial area around the throat section, where erosion always has an impact on accuracy.

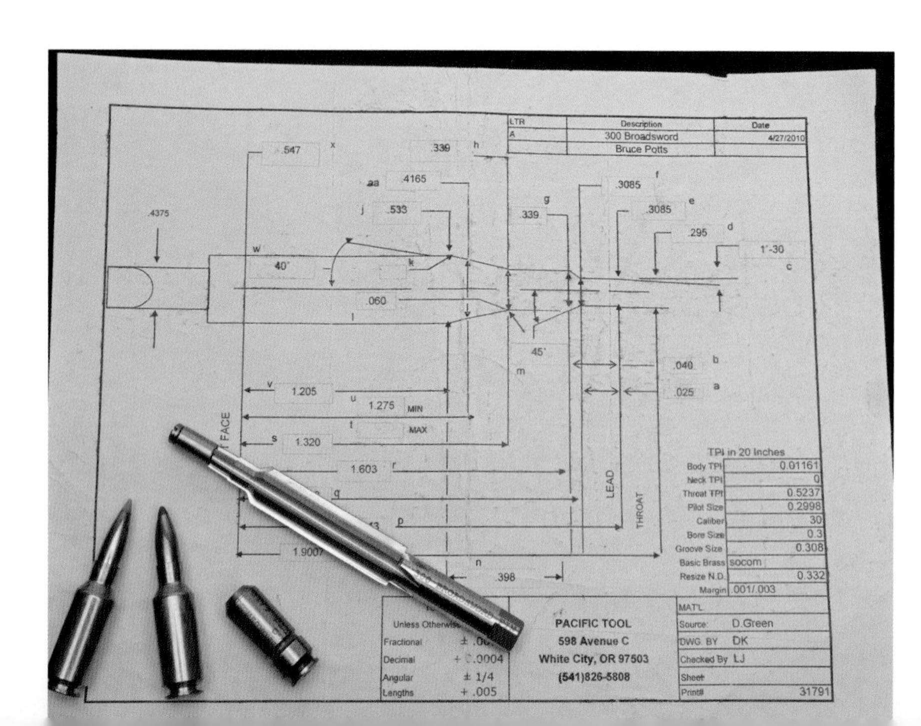

The QuickLOAD ballistics program helps in the design of new reamer prints to ascertain the optimal neck length, diameter and capacity of any new rounds.

Shoulder Angles

This is another important area to consider when choosing a reamer. The angle of a shoulder can influence the smooth loading of a cartridge through the magazine: here a shallow angle means smooth loading, whereas sharp angles can cause hang-ups. The main reason to change the shoulder angle on a case is to increase the overall powder capacity, for example by altering an angle from 18 degrees to 40 degrees. Other concerns such as efficient powder burn or gas flow around the case can be greatly influenced by the shoulder angle. The intention is to burn the powder efficiently and direct all that energy at the base of the bullet, not at its sides. The 5mm × 35 Smc cartridge from Mic Mc Pherson has a radiused shoulder to eliminate this problem and takes cartridge design to a new level.

Leade or Throat

This is the area just in front of the case neck in the barrel's chamber. It is the bullet's first contact with the bore and the way it enters, its orientation, pressure, angle and diameter are crucial to the bullet's eventual passage down the barrel. The leade on a factory rifle is standard to a particular cartridge and takes no account of the actual bullet weight or profile you wish to shoot. The diameter of a throat should be the minimal diameter for that calibre bullet. If it is too large the bullet can start to wobble in the throat before it becomes fully engaged in the rifling, so compromising accuracy.

The throat angle is very important in relation to bullet stability, rotational twist and wear to the initial part of the rifling caused by heat and erosion from hot gases. Ideally the throat angle should try to match the angle at which the bullet actually touches the bore's rifling lands.

This is determined by measuring the diameter of the bore where the bullet's ogive touches the rifling and then where the rear bearing surface diameter of the bullet touches. The angle between the two when a line is drawn is the optimum leade angle. This will differ depending on the ogive of the bullet used, which is why some bullets shoot better than others in certain rifled barrels: 1.5 degrees is the standard for .243 and .308, 3.0 degrees works well for .222

Matching a barrel's chamber to a specific bullet or the range of bullets you intend to use can be a time-consuming process, but it is essential part of any custom project to achieve the best results.

and 1.0 degrees is good for .270. This will obviously change in response to wear from erosion as the round count of bullets fired increases.

Factory SAAMI Specification

Most reamers are made to the industry-standard SAAMI (Sporting Arms and Ammunition Manufacturers Institute) specification, but that is intended for the dimensions of an average chamber. If you are going for a custom rifle I

Reamers have a hard life: Paddy Dane uses a Red Elliot stoning fixture to sharpen and modify his reamers.

highly recommend spending a little more on a custom reamer to chamber a barrel for the cartridge you intend to shoot. A specialist tool such as the Red Elliot stoning fixture allows you to resharpen and profile a reamer on the bench and is the tool of a real gunsmith.

WILDCATS

Wildcat calibres, by their nature, do not have standard dimensions and so each individual reamer is specific to that cartridge. I use the QuickDESIGN and QuickLOAD ballistics software to design my cartridges and reamers. It allows you to tweak and readjust measurements until you are happy with the results, and then run them to test the theoretical ability of the new cartridge in a simulated environment. You can instantly see any flaws or advantages against known cartridges in the database before you commit to a reamer maker and order the wrong reamer.

Half the fun in devising your own wildcats comes from changing the shoulder angles to gain more powder or from increasing the neck length dimensions, which can be used to improve the bullet grip and stabilization in the case or extend the neck to allow the bullet to sit on top of the powder without compressing it. The neck dimensions and throating angles on any reamer are crucial to maximizing the case performance and still maintaining safe pressure levels.

With nearly all my wildcats I choose a piloted reamer. This is because you will find that if, for example, the barrel is a .308 calibre, the groove diameter will be 0.308in while the bore diameter can be 0.2971in to 0.3000in. When you order the reamer, however, the rotating bush for the piloted reamer can be either too tight or too loose. Manufacturers' barrels or tolerances can vary a lot. This is where a good custom riflesmith can turn up a new pilot to fit your custom barrel exactly, ensuring that the reamer is central to the bore and the chamber is completely concentric. A sloppy reamer pilot will otherwise allow a skewed or wobbly reamer as it cuts. I have found that a carbide reamer cuts smoothly and is worth the extra expense, even if it is used only for one chambering and usually as a finisher reamer (Steve Bowers, for example, first bores out the excess steel by hand).

A wildcat is often only a slight redesign of the original parent's case. Here a standard 7mm Mauser round is seen next to a 7mm Mauser Ackley Improved round, which has straighter case walls and sharper shoulder angles for more powder capacity.

I also nearly always order a reamer without a throat: a separate throating reamer using the same pilot bush as the finisher reamer can be used to cut a throat that will accommodate any bullet style and weight, so you do not have to order a completely new reamer if you change your mind during the build. This is very handy for long bullets in small cases where you want to seat them out further to maximize powder capacity, for example with .224 calibre 90 grain bullets in the .223 Rem or .22 Satan. Another example may be an over-bore capacity case like the .20 Satan, where you will never fill the case and so you can seat the bullet further in the case and down the neck. This means that if you throat that chamber with the minimal throat length, you can progressively reseat the bullet as it wears out the barrel. In other words, you are chasing the rifling up the barrel but you can do this because you have a short throat in the chamber and the bullet still has enough neck tension or grip to hold it securely and release positively.

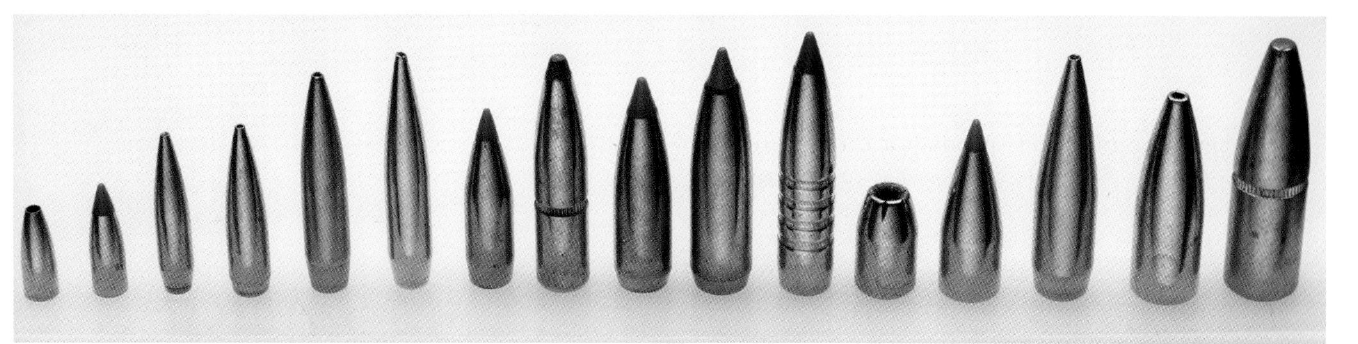

Due to the vast array of bullet types, weights and calibre sizes available, it is essential that the rifling twist of the barrel is correctly matched to ensure stability and accuracy.

RIFLING TWISTS

Choosing the correct rifling twist rate causes more problems than any other aspect of a custom rifle build. The twist rate of rifling can be expressed as, for example, 1 in 10in, where one complete turn of 360 degrees is performed within the length of the twist rate. Bullets of differing calibres, lengths and weights require a different twist rate to stabilize the flight and you need to select the best for your choice of bullets. Bullet choice for one particular calibre, however, can be large. The .223 round, for example, offers choices from a 30

Calibre	Common twist rate	Alternative twist rates	Specialist
.14 (.144)	8.75	9 or 10	n/a
.17 (.172)	9 or 10	7, 8 or 12	6 for 37gr VLD bullets
.19 (.198)	13	11, 12	9 for heavy bullets
.20 (.204)	10 or 12	14 or 15 for light bullets, 8 to 9 for heavy bullets	7 for specialist heavy bullets
.22 (.224)	10, 12 or 14	7–9 for heavy bullets or 15–17 for light bullets	6 for heaviest weights
.22 rimfire (.222)	16	9–14 or 17–20	6–8 for heavy bullets, e.g. SSS Aquila 60 gr
.243 (6mm)	9 or 10	12–16 for lighter bullets	5, 6 or 7 for 100gr plus bullets
.25 (.257)	10	11–14	9 for heaviest
.264 (6.5mm)	8 or 9	10, 11 or 12 for lighter bullets	7 heaviest of 160gr plus
.270 (.277)	10	12 or 13 for 130gr below	8 for heaviest
.284 (7mm)	9 or 10	11–15 for less pressure lighter bullets	7–8 for heaviest of 175gr plus
.308 10 or 12		8, 9 heavy or 13–18 for lighter-weight bullets, max velocity	4–7 for subsonic work with heavy bullets of 220gr plus
.303 (.311)	10		Limited bullet choice
.323 (8mm)	10	9 or 12	n/a
.338	10	11–14 for lighter bullets 180gr	6–9 for heavy bullets and subsonic work
.35 (.358)	12 or 14	10 or 15 for bias toward light or heavy bullets	n/a
.366 (9.3mm)	12	10	n/a
.375	12 or 14	15–18	8–10
.411 (.410)	16	20–24 lead bullets	n/a
.416	14	10–12	8
.458	14 copper jacketed or 20 lead bullets	15–26 for lighter jacketed bullets or lighter lead bullets travelling fast	8–12 for heavy or subsonic loads
.510	18	14 or 20	9 for subsonic loads

grain Barnes all the way up to a 90 grain JLK or Sierra Match Kings, but the trouble is that one twist rate will either stabilize the heavier bullets or over-rotate the lighter bullets. A 1 in 14in twist is fine for the lighter bullets as it allows higher velocities without too much rotational torque, which a faster twist barrel would cause, while keeping the pressure levels within safe limits. A heavy 90 grain bullet, however, will need a 1 in 6.5in twist to stabilize and so more rotational spin, but this would tear a lighter bullet apart.

When a bullet enters the bore as in the case of a lead bullet or jacketed lead core, it is squeezed to size and expanded or obturated to fill the groove space, so allowing the gases to work efficiently on the bullet's base. Once the bullet is locked into the rifling, it is very important that, as with the bore dimensions, the rifling twist does not change as this will affect accuracy. As well as the correct twist rate for the bullet weight, it has to be totally uniform along its whole length. This is why cut rifling barrels are popular, whereas some cheaper button-rifled barrels can have small variations in the barrel twist rate. A decreasing twist rate along the barrel length is worse than a slightly increasing twist rate, because the rifling will cut a wider groove with a decreasing rifling twist rate. As the bullet rotates it can therefore allow the bullet to yaw around its axis and cause larger groups downrange.

Monolithic (solid) copper bullets have limited expansion, which is why they need special seating depths. Seating these bullets well off the rifling lands, perhaps at a distance of 50–60 thou, allows the bullet a degree of travel before it engages the rifling, giving it better potential accuracy and lower initial pressure.

CHAMBERING A BARREL

Many gunsmiths have their own way of chambering a barrel. This is how the custom rifle maker and precision engineer Steve Bowers does it in eleven stages.

1. Profiling a barrel blank to a custom profile for the barrel tenon.

2. Truing up the barrel blank to the bore axis for perfect concentricity, ready for screw cutting and chambering.

3. Truing spigots, these are ground between centres to each barrel's unique bore dimensions.

4. Screw cutting the barrel tenon to match the action/receiver.

5. Boring the barrel recess for the bolt shroud in front of the lugs for a Remington-style bolt.

6. Roughing out the chamber with a twist drill instead of a roughing reamer.

7. Boring the draft angle of the chamber to true up from 5 to 8 thousandths of an inch off the finished chamber size.

8. Using a finisher reamer to bore the final dimensions of the chamber.

9. Using go and no go gauges to check the correct headspace.

10. Fitting the action to the barrel to check that the bolt locks and that the headspace on the gauges is correct in situ.

11. Muzzle threading and crowning to finish.

FULLY SILENCED BARRELS

These barrel types are highly specialized and incorporate a full-length suppressor system. They can be very popular on a hunting rifle where overall length needs to be kept to a minimum and noise reduction is paramount. The trouble is they need specialist barrel profiling, porting and screw cutting, which adds to the cost and the skill level necessary from your custom gunsmith. I use two fully suppressed rifles, one a .22 rimfire and the other a .308 Winchester. These were actually Sako custom items for specialized use only. The .22 rimfire has a ported barrel just in front of the chamber to bleed pressure that is diverted into a sealed outer chamber. As the bullet exits the muzzle, the remaining gases are then passed through a series of baffles, resulting in only the firing pin noise being heard. A modern equivalent is the MAE Phantom, which uses a fully silenced shroud for a Sako or similar .22lr rifle. It is phenomenally quiet and accurate, too, making it the ultimate vermin tool.

The fully silenced MAE Phantom rifle is the quietest .22LR rifle you will ever shoot and offers bullet on bullet accuracy.

MAE fully silenced full-bore 6.5 × 47 Lapua deer rifle is both very quiet and very accurate. As such it is best in its class.

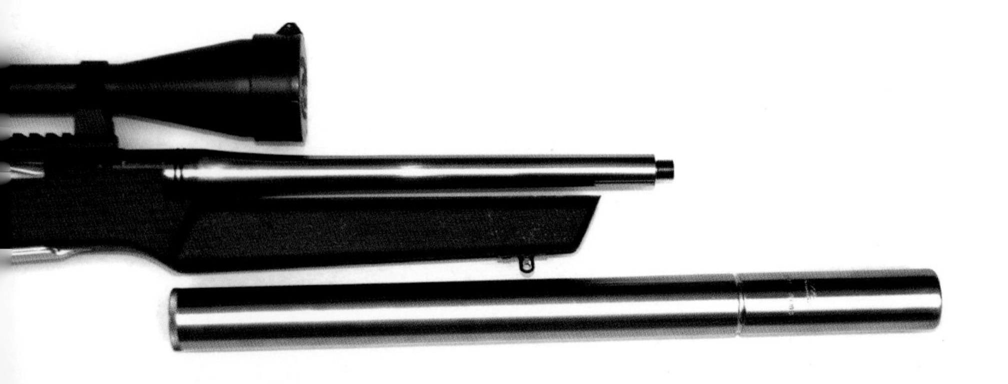

Removing the MAE silencer reveals the short barrel beneath the stainless steel shroud, sealed with a double 'O' ring near the action end.

CUSTOM MUZZLE CROWNS

The end of the barrel is actually probably the most crucial area. It is here that the bullet leaves the confines of the barrel, which has steered the rifling and stabilized its environment, to enter free flight on its own. Any irregularities that leave it out of true, or any damage, corrosion or flaws in this area, can greatly affect those crucial milliseconds as the bullet jacket parts company with the rifling. Any defect will have an effect as the bullet exits and ultimately ruin its accuracy downrange. Tightness or choke can be a good thing, however, because it is said to regularize an otherwise average barrel. It is difficult to qualify in reality, but when using .22 rimfire some form of choke or tighter muzzle end is advantageous. What you are striving for is the uniform exit of the bullet in perfect rotational flight, in which case the kinetic energy provided as a push on its base by the gases from the burnt powder charge does not escape around the bullet at odd angles as it leaves. If this is not achieved it will make a bullet yaw from the muzzle. Characteristic soot or burn marks left on the muzzle crown can indicate a perfect muzzle exit with a uniform star shape of equal dimensions. Any yawing is shown as uneven stars or as missing points. Another benefit of a good muzzle crown is that it protects the rifling emerging from the barrel. In a recessed or deep concave crown the rifling sits below the muzzle edge and offers some protection from knocks and scrapes.

The muzzle crown is the last contact that the bullet has with the rifle before free flight, and as such it needs to be perfect: (left to right) 11 degrees, flat faced and recessed.

Types

As you would expect, there are many different types of muzzle crown available to the custom rifle maker. Typical crown types are cut on the lathe when the barrel has been set up in a position concentric to the bore axis and not the outside of the barrel dimensions. By cutting the crown you are eliminating any irregularities and making sure the rifling ends are all identical and perpendicular to the bore on exit.

Concave

In this type there is a swamp to the crown or, in other words, that from the rifling down to the edge of the barrel there is a degree of profile, ordinarily 11 degrees. This is often called a target crown as it allows gases behind the bullet, as it exits the barrel, to smoothly pare away along the scallop and not interfere with the bullet's flight.

Flat

There is nothing wrong with a perfectly flat muzzle crown. When done well they allow a very good exit for the bullet, but you have to be careful not to knock the end as the rifling is more prone to damage without some form of recess to the crown.

Recessed

The muzzle has a stepped profile where the bore exit is well below the actual end of the barrel. It is cut at a 90-degree angle and usually has a 40–50 per cent edge before it steps down to the bore. This allows maximum protection from any damage occurring to the rifling.

Chapter 4
Scope Mounts and Sights

This is where everything starts to come together and it is vitally important to get the sights correct. While correct fitment is essential for an accurate shot, their placement, size, shape, colour and fitment also affect the way a custom rifle handles, and very importantly, looks. Most, if not all, new custom rifles are built to accommodate some form of scope, which is the most popular choice for the modern shooter. Traditional shooters and hunters, however, still like the look and reliable function of a good set of open or iron sights. Many custom rifles are deliberately ordered with open sights as the main choice of sighting as there is still nothing more reassuringly reliable for fast, close-quarter shooting. There is no reason why a scope or open sights cannot be fitted to the barrel, as mistakes or accidents happen in transit and nothing is more frustrating than turning up in a hunting camp and finding your scope damaged beyond repair. A back-up scope or set of open sights really can save the day.

This is equally true of the mounting system, which is the only thing between the scope's aiming reticule and the rifle that propels your bullet. If you get all the other factors right in a custom rifle, but neglect to spend time on a decent set of mounts, all the hard work will be ruined. The type of scope used is also very important. I have seen many beautifully proportioned custom rifles ruined by an excessively large or heavy scope fitted just because it cost the most. Custom rifles should be all about proportions and elegance, but there is no accounting for taste. This chapter will run through both open sights and scope mounts, fixed and detachable, along with some interesting scopes and scope fitments for the modern custom rifles.

Scopes and mounts have come a long way, but both have a place in today's custom market.

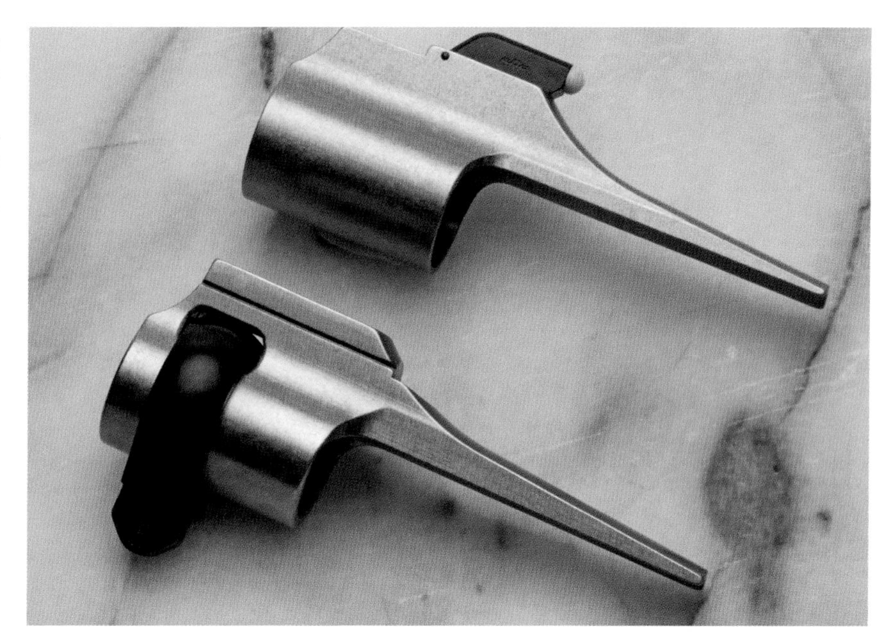

Recknagel offers excellent custom English-styled foresights, such as this H&H style and banded fibre optic sight.

OPEN SIGHTS

Often forgotten on modern rifles, open sights still have a vital role to play, not just aesthetically on a classic design, but they could save your life. Open sights consist of a foresight, fitted with some form of blade or bead, that aligns with a rear sight, which is adjustable for elevation and windage. The refinement comes in the degree of finesse applied between looks, function and reliability. Open sights can be simple or complex, but the design should reflect the duties necessary for the rifle they are mounted on. Rather than an afterthought, open sights add a degree of elegance that easily surpasses any telescopic sight. The huge variety of open sights available include standard factory fitments, refurbished stock items, specials custom-made from scratch and all manner of aftermarket options.

Classic

Classic open sights traditionally refer to barrel banded foresights, ramp sights and folding rear sights with extra leaves and quarter ribs. This is a very English style that works well on any classic custom rifle whether you mount a scope or not. There is something reassuringly British about a classic set of open sights, which look just right on a blued barrel actioned rifle project. Foresights usually have a ramp feature, but some have the addition of a barrel band to fit perfectly around the muzzle's circumference. These can be made to measure, which is expensive, but there are many aftermarket sources: Recknagel, for example, has a superb range offered in white so your custom gunsmith can produce your preferred finish. They can be attached by soft solder or pinned.

On a custom rifle project you are able to order exactly what you want, including a choice of custom foresight elements. Straightforward blade or serrated finished blades or barleycorn-type posts are fine, but I spend the extra and go for some form of adjustable foresight. This can have height adjustment to meet the trajectory of differing bullets, windage adjustment or replacement sighting elements. The Recknagel Holland and Holland type foresight has a small and precise aiming bead, often made from ivory, gold or silver, that is fine for daylight hours, but some form of highly visible sighting has been found to be necessary in the past when, for example, shooting leopards after dark,. One simple yet effective alternative is a flip-up cover that shields the original bead, presenting a larger and brighter ivory sighting bead.

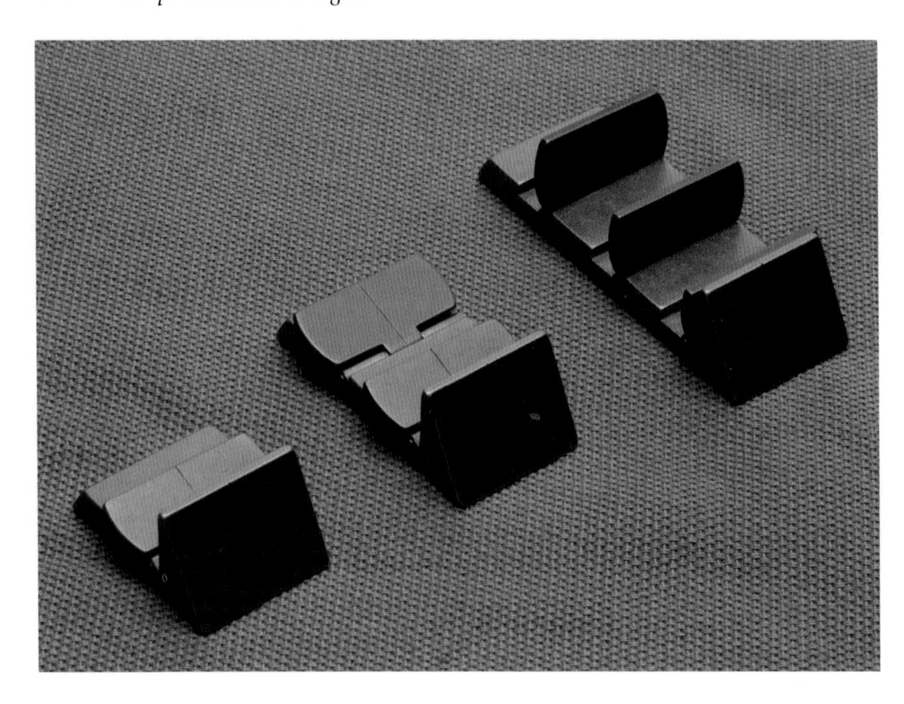

Flip-up rear sights of one, two or more leaf construction are traditional sighting for any new custom rifle. These are from Recknagel.

Rear sights on many classic sporting custom rifles used to have quarter ribs or half ribs that were used to attach a series of folding rear sight elements. Arranged as a fixed element for 100 yards or closer, depending on how the rifle or ammunition was sighted, there is then a series of one or several folding leaves that can be sprung up to compensate for a bullet's trajectory downrange, for example for 200, 300 or 400 yards. The beauty of a quarter rib is that it looks very elegant and lends itself to some form of custom engraving or embellishment.

Modern

Fibre optics have become very popular on sporting arms, especially on true hunting custom rifles that need to earn their living. The fast target acquisition offered by bright fibre optics is a bonus. Again Recknagel has a huge range of fore- and rear sights of this type, including ramp type or soldered-on banded ramp models. Each has a blade inset that can be matched or altered to the hunting, targeting or running game scenario, as appropriate.

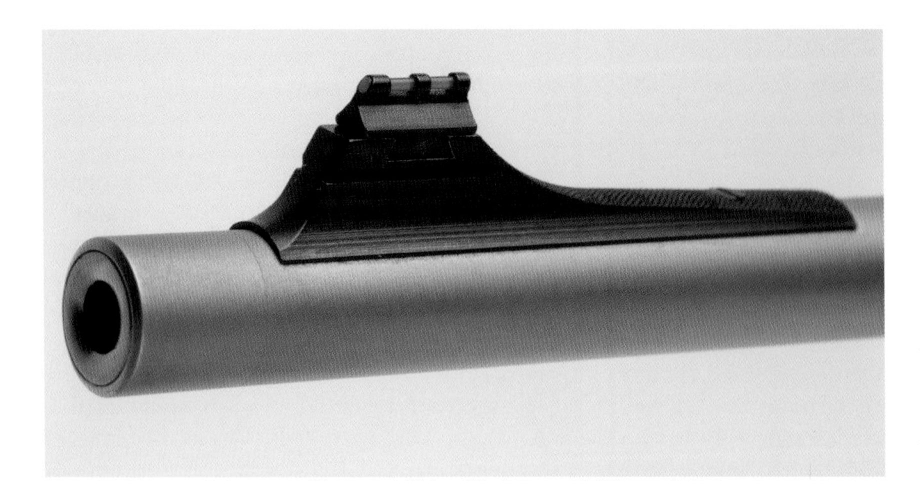

Fibre optic sights are a valuable addition to modern day custom rifles as the target is clearly visible.

Recknagel offers a superb range of fibre optics for fore- and rear sight arrangements, here seen on a barrel banded rear sight custom attachment.

These are very easily exchanged and the range of sizes and styles will match any project from classic to ultra-modern tactical.

Ghost Sights

These are very popular on large-calibre and big game sporting rifles due to their fast target acquisition, wjich allows an accurate shot at close range. As the name suggests, the sights are made so as not to interfere with the sight picture, appearing as a ghost image around the target with the foresight in sharp focus. The rear section is usually a large ghost ring of large aperture that allows the foresight bead, blade or an exotic insert such as ivory to centre very quickly and naturally within the ring. They are usually fixed at a close range, such as 60 yards, although some adjustments for windage and elevation can be made. Alternatively the ghost ring can be tilted forward to an alternate aperture size.

If the custom rifle has a Weaver-type scope mount fixture, a ghost ring sight can be easily attached and removed as the occasion arises. I remember shooting a 550 magnum rifle in South Africa based on a straight-cased 460 Weatherby magnum cartridge. I was glad it had ghost sights fitted otherwise I would have

Ghost sights can be a lifesaver on heavy recoiling custom rifles where an instant shot is needed with both eyes open.

Ghost sight fitted to Professional Hunters Jason van Aarde 550 magnum rifle.

finished up wearing it due to the excessive recoil.

The Recknagel ghost/peep sight, available as a fixed, detachable, rail or Weaver base attachment, is a typical example because of its quick, detachable facility and larger than normal aperture for fast target acquisition and foresight alignment.

BELOW: *The most universally used mounting system today is the Weaver type mount, which allows a swift scope change from rifle to rifle.*

SCOPE MOUNTS

Scope choice is personal, but getting it just right while not spoiling a custom rifle's balance in both looks and handling is quite tricky. A smaller scope is often better if you are putting together a classic lightweight sporting rifle, whereas the larger long-range rigs can cope well with those huge and expensive European scopes that look so impressive.

Weaver, Picatinny

A Weaver or Picatinny-type mounting system makes real sense for the many synthetic stock semi-tactical rifles now available. They are not

A Weaver one-piece mount is handy for mounting night vision equipment or scopes with non-standard body lengths.

the prettiest of mounting systems, which is why they seldom appear on classic rifles, but their universal fitment make them very popular and they are available in an amazing array of heights, MOA compensations, steels and tactical rail additions.

MOA Adjustments

Scope mounting rails that allow a certain degree of trajectory compensation are now popular as increasing numbers of custom rifles are built for long-range shooting.

A scope mount has a tilt that allows your existing scope to use more of its internal adjustment for elevation. This is because the mount is angled forward with the scope pointing down towards the barrel, hence you have to elevate the muzzle more to sight your target. You can then set your scope at its lowest elevation setting to zero, leaving the rest of the elevation adjustment available for trajectory compensation. The mounts can be in one or two pieces, and can also be separate items or machined into the top of the action, usually with a universal Weaver-type attachment. They are commonly sold as MOA (Minute of Angle) compensated, referring to the one-minute angle of arc adjustment from the horizontal. The common 20 MOA rails, for example, allow twenty more inches of range at 100 yards, 40 MOA allows forty more inches and so on. You can buy an MOA compensating

Scope mounts with a built-in slant angle downwards, such as this Ken Farrell 20 MOA mount, allow increased range adjustment from your scope downrange.

Custom MOA mounts, such as the Recknagel 70 MOA adjustable mount, allow a huge minute of angle adjustment for long range while maintaining precise accuracy.

mount with 0, 5, 10, 20 or 30 MOA adjustment to match the cartridge and range you intend to shoot at. One very interesting MOA adjustable scope mount is the Recknagel Era Tac adjustable inclination mount, which allows an adjustment range from 0 to 70 MOA. It is a one-piece mount available in three ring sizes and differing heights for scope objective lenses and fitment may be either by locking nut or quick detachable. The two securing cross bolts are loosened and this allows the mount to pivot from the solid tapered front joint by rotating the rear cammed thumbwheel marked in 10 MOA increments. It is very precise and locks up solidly for really long-range precision shooting potential.

SCOPE RINGS

Integral

Either the existing scope rails can be used or tapped receivers can be fitted for matched or aftermarket generic mounts. Manufacturers often machine scope mounts so that they accept only their own mounts or include a dovetail rail for their own mounting systems. Others supply their rifle actions drilled and tapped for a set of bases that either mount a dedicated scope mount or allow a generic mount, such as the Weaver or Picatinny mount, to be fitted, thus allowing a far larger range of scope rings to be fitted. Neither system is any better than the other, although integral dovetails in the action top allow a one-piece scope mount to be fitted, whereas a Weaver type or Picatinny requires a base and separate scope rings to complete the scope mounting process.

Sako and Tikka Optilock

Although these are standard factory mounts, they offer excellent quality and value for anyone who requires a solid fitment. The stainless and blued finish blends well with any action type and, although primarily designed to fit Sako and Tikka rifle scope rails, the fact they can be bought as a one- or two-piece allows you to choose a Weaver-type mounting system option, which makes it even more versatile. The

Sako / Tikka OptiLock mounts are still a good choice for custom projects as they are very well made.

Optilock has inserts that secure the scope into the rings and will not mark or crush the scope, while also allowing a certain amount of tilt for correct alignment. Extra adjustment is possible by packing or shimming the rear mount to tilt forward as a semi-MOA base configuration.

Leupold

Leupold makes scope mounts that are well suited to both classic rifles and modern styles. Their standard series of rings and bases come in one- and two-piece configurations and are made from steel that is available in blued, gloss, matt or silver finishes to match your rifle's action and barrel. They can also be engraved easily and colour case hardened. The standard series has a front dovetail ring and rear windage adjustable ring, making them practical as well as elegant. The bases are sculptured and a two-piece base set allows good access to the bolt opening and breech area. One-piece bases that use the same standard ring sets have a dovetailed front section, an integral bridge for strength and a rear section with windage adjustment screws that also secure the rear rings.

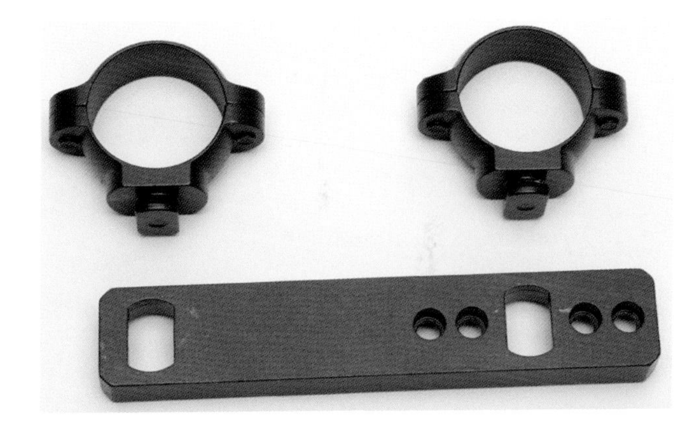

Leupold dovetailed scope mounts are made from steel and so are suitable for rifles requiring a classic look.

They are available in three finishes and fit a wide range of action types. A double dovetail fitment may also be supplied.

Tier One

Based in Yorkshire, this division of Evo Leisure Ltd offers fixed mounts that are pre-

Tier One scope mounts and rails, which are made in Britain, offer a wide range of ring sizes and mounting options for any custom project.

cision made on multi-axis CNC machinery from 7075T6 aluminium. They are available in a various heights, can fit 1in (with spacers), 30mm, 34mm and 35mm scope body tubes, and are hard anodized for longevity. They are very well made and finished with perfectly matched top and bottom ring halves for a uniform support of the scope tube, so there is no crushing due to the scalloped sides. They are numbered to each other so you can locate the top part the right way round if you remove a scope. An integral cross slot provides a precise location to a Weaver base and ultimately stops movement under recoil. The four Allen screws exert an even pressure. The one-piece mounts can have a built-in MOA compensation for maximum elevation adjustment and extended long range, or just accurately centring the scope optically.

Burris Signature 30mm Zee rings

These very clever scope mounts solve the age-old problem of lack of elevation and windage adjustment. They use two separate inner rings that cradle the scope to stop slippage or damage to the scope body. These are called Zee rings and fit the Weaver-type base, but standard mounting arrangements using a single- or double-dovetail arrangement for specific rifle actions are available. The eccentric polymer rings come in two halves or shells that are numbered 0, 5, 10 and 20 in both negative and positive values, so indicating the amount of tilt or correction in one thousandths of an inch. The 30mm mounts come with 10 thou inserts and are tightened by Torx head screws. The mounts are available in gloss and matt blue or nickel. If you put a +10 in the rear mount and a –10 in the front

Burris Zee rings have a unique cammed plastic insert to allow adjustment for elevation and windage.

mount the scope will tilt down and therefore give you an additional elevation adjustment of about 20in, depending on the mount spacing, without having to use the internal adjustment in your scope.

Talley

Talley Manufacturing in South Carolina has been producing well-made and attractive scope mounts for custom rifle projects for more than forty years. They offer a comprehensive range of mounts and mounting systems, including Talley Lightweight alloy mounts for lightweight stalking rifles and fixed ring systems, for example for 6mm PPC Sako or Tactical Picatinny rails. The quick-release rings come in 1in, 26mm, 30mm and 34mm, rimfire and speciality options. Of special interest to custom rifle makers is the choice of gunsmith blank bases, barrel bands, inset sling swivels, custom bolt handles and cross bolts.

Talley mounts are steel and both well made and good looking. They take a traditional blued finish well.

Conetrol projectionless scope mounts have a streamlined design and are superbly made. They will complement any custom rifle project.

Conetrol

Conetrol offers truly affordable and good-looking scope mounts notable for their sleek appearance without any unnecessary or unsightly scope screws. The use of a projectionless split-ring mount cut from solid ensures that torque cannot be inflicted on the scope tube or receiver. It has the strongest windage mount design because the dual-ring movement to both scope mounts, front and rear, allows the scope to be centred on the gun receiver after adjustment. These beautifully made mounts look superb on a blued steel, fancy walnut-stocked custom rifle. Both rings and bases are detachable and are available in three different types. There come in five ring sizes and numerous height options. Conetrol mounts are available in gloss or matt bluing and stainless, or in white for custom smiths to finish as they choose.

QUICK RELEASE

On a custom rifle that might perform a double duty as a vermin/deer gun or deer and large game rifle, it really makes sense to fit a quickly detachable scope mount. This allows for using a low-powered scope for all-round use and then quickly changing it for a longer range or better light-gathering scope.

EAW Apel Pivot

Ernst Apel Würzburg (EAW), founded in 1919, has a well-earned reputation for quality scope mounts that fit an enormous range of rifle actions. All mounts are produced on their premises in Gerbrunn, Bavaria, to maintain the highest standards. Although EAW produces claw and roll-off mounts, the company is best known for its swing-off or pivoting mounts. These come in two parts, the base that

Apel pivot or swing-off mounts from Germany are expensive, but offer a return-to-zero scope mounting system that can be quickly detached.

fits directly to the rifle's action and then the separate top rings that fit scopes from 1in up to 34mm. There are also rail-type mounts, and EAW's products can produce two thousand combinations.

EAW swing-off or pivot mounts allow the scope to be taken on and off quickly and, more importantly, precisely. The front ring has a dovetail that fits into an identical union in the base mount, but it locates at a 90-degree angle and then the mount and scope is swung inward onto the rifle. Protrusions on the rear mount push a camming locking lug that then snaps shut, totally rock solid. A small ball-ended lever to the right of the base can be raised and the scope then swung off the rifle to the right in one stroke. Adjusting screws to both bases and top rings allow perfect alignment as well to minimizing stress to the scope body and ensuring a tight fit.

Ziegler ZP Contra Mount

Ziegler Präzisionsteile is a relatively new name that designs and develops precision-engineered components using state-of-the-art CNC machining. Ziegler scope mounts employ a claw-type mounting system in which hooks or claws from the top rings fit precisely into the mount bases on the rifle to achieve a positive union with no zero loss. These ZP Claw mounts are available in the tip-forward, quick-detachable mode, but on the Contra claw mounts the scope is detached by tipping the scope rearward. The Contra mounts come in three parts that require an exacting fitment. They are extremely precise and allow a very low mounting height to your rifle. A large objective-sized scope can be attached and reattached without loss of zero. The claws on the rings fit so that when the rifle recoils the claws grip their bases more tightly and do not move up, unlike the older-style claw mounts. To re-

The Ziegler ZP Contra Mount is a return-to-zero scope mount with a super-fast release system.

The Recknagel Era swing-off mount is available in an amazing array of styles and fitments. This G9-Plus mount has a unique rotary locking system that cannot work loose.

move the front ring, slide back the twin knurled levers on the side of the base, which releases the claws from their precision fit. Ziegler ZP Contra mounts are expensive but they look superb, adding little height, and they retain your rifle's zero when attaching and reattaching.

Recknagel Era Swing Mounts

The comprehensive Recknagel Era range of swing on/off mounts is another that allows instant scope removal or replacement in the field, for storage or travel. It is available in a wide range of scope ring sizes, scope heights and styles. The G9-Plus range is a steel two-mount system that provides a low scope mounting to the rifle for good looks and ergonomics, while providing exact zero repeatability when removed and reattached. The front ring rotates in at a 90-degree angle to lock in the mortised tenon of the mount as the rear mount is released. This rear mount has a rotary locking system that makes it impossible to rotate or loosen the scope when locked in place. There is also a levelling feature that maintains a proper scope alignment with the bore of the gun so you do not have to shim the bases.

Leupold QRW

The QRW or Quick Release Weaver provides a quickly detachable mount for fitting to a universal Weaver base, although mounts for many other bases or rifle actions are available with

These Leupold QRW mounts on a Norman Clark 458 Lott custom rifle allow the scope to be removed quickly so that a ghost or red dot sight can be added for close target acquisition.

the standard series scope rings and bases. The QRW series are available in a number of finishes from gloss and matt to silver, with scope height options for 1in or 30mm mounts. There is a small adjustable lever to the right side that tightens and releases the mount from the base, and the cross-slotted lug stops the scope moving and helps position it back to zero again. QRW mounts use Torx head screws to secure the scope in place so it is gripped evenly and tightly to avoid slippage. The attractive mounts have a smooth rounded top with cutaways where the top ring touches the bottom mount halves so as not to pinch your expensive scope.

Warne Maxima Q/D

Warne, based in Oregon, produces a wide range of scope mounts to fit most action types with fixed and quickly detachable options and scope sizes from 1in, 30mm and 34mm. The Maxima Quick Detachable (Q/D) mounts are one-piece construction from a sintered steel operation to provide strength and precisely made parts.

Warne Maxima Q/D mounts are another scope mounting system. Many of this type are action dedicated and I use a set on my .35 Whelen AK Improved Tikka rifle.

Smithson custom scope mounts are of exceptional quality and grace some of the world's best custom rifles.

They can be finished in powder-coated matt or gloss, or nickel electro-plated, with a variety of heights to suit differing sized objective lens scopes. The Q/D release lever is mounted separately and the handle can be adjusted independently to angle it as you like. Because the Torx screws are located with two at the top and two at the base, the top part of the mount has a humped look. Beneath the scope base there is separate recoil or location lug that sits into the Weaver-style base. This helps relocate the mount to position it correctly and stops scope movement under heavy recoil.

Smithson's

Joseph Smithson has been a full-time gun maker since 1985 and has run his own shop in Utah for twenty years. Smithson is known for his custom rifles in bolt actions, single shots, double rifles and shotguns, but in Britain his reputation is based on his outstanding quick-detachable scope mounting systems for classic-looking quality rifles. A male dovetail assembly on the underside of the scope rings fits precisely into female dovetails on the scope mounts. These cuts are made directly into the double bridge-type integral action top or on separate bases. Either way the mounts are released by a single side-mounted push button that allows the scope to unlock and relock with no zero loss as a large ball bearing maintains a tensioned lock-up when secured. With the scope removed a ghost or peep sight can be inserted in the rear base or receiver cut dovetail, making for a fast, close-quarter, dangerous game gun.

One- offs

These belong to the realm of the true custom makers. Range-adjusting mounts made from scratch are not cheap and require the best precision engineers.

A mount that always turns heads is the .338 BR subsonic mount made for me by Steve Bowers. It provides trajectory compensation allowing hits subsonically at 500 yards. The mount has a 120 MOA bias, so ballistics downrange are a 250 grain Lapua Scenar bullet with a 0.675 ballistic coefficient and 1089 fps, which still has a 933 fps velocity at 500 yards, only a loss of 156 fps. The energy from the start is 658 ft/lb and is still 483 ft/lb at 500 yards, but the bullet's trajectory has dropped by 328.9in at that range, hence the need for a large MOA scope mount. It is notable, though, that in a 10mph wind it drifts only 16in.

The custom scope mount built by Steve Bowers for 500 yard subsonic use with my .338 BR (Whisper) rifle is crazy but fun.

SPECIALIST SCOPES AND SIGHTS

Barrel-mounted Scopes

These may be an older design, but I still love the look of older traditional varmint rifles. With the increasing interest in classic rifles, as with classic car collectors, there is a demand for retro-looking rifles and their associated scopes and mounts. Collectors just have to own period pieces and you can have a custom rifle specifically made with modern materials but with that traditional look.

I have a lovely Tikka M65 custom rifle built by Ivan Hancock of Venom Arms with a 1 in 14in, twist match-grade, stainless steel Pac-Nor 28in barrel length that would benefit from the Swift's characteristics in powder burning abilities. In fact the whole Tikka M65 action once originally belonged to a .270 estate rifle sourced from the gunsmiths R. Mcleod & Son in Tain, Ross-shire. The complete package was concluded with a vintage, by today's standards, Tasco pre-1970 barrel-mounted scope some 28in long and possessing a fixed 16× magnification. It has a recoil spring, external windage and elevation adjustments dials so characteristic of the era of classic varmint rifles.

Other vintage scope makers include Unertl, Fecker or Remington. One of these allows you to create a custom build rifle that mimics the char-

I still love the old-fashioned barrel-mounted scopes. They make a fine scope to add to a retro custom rifle, such as this Venom Arms .220 Swift on a Tikka M65 action.

Another classic scope mounted to a Remington 40X .22 rimfire for 200 yards shooting. This Tasco vintage 1971 scope is a copy of the Unertl American scope.

acteristics of the past with a classic rifle or placed on a modern custom. I like the way these scopes make you take your time and think about how you shoot, so when you connect using this vintage glassware the shot seems that bit more special.

ABOVE: *Custom scope mounts take many forms. A Kershaw MAE silenced fox rifle is here fitted with a Hensoldt night vision forward-mounted scope to the AX stock rails.*

LONG-RANGE SCOPES, LASER RANGEFINDERS, NIGHT VISION AND THERMAL IMAGERS

The growing demand for increasingly elaborate or specific custom rifles has led to some extraordinary looking and performing rifles. A big part of British shooting involves night shooting for vermin. Many custom rifles are designed for the attachment of night sights, lasers, illuminators or thermal imaging devices. Some of the custom rifles I have used with these sighting systems are illustrated here.

RIGHT: *Dedicated Pulsar NV 870 night vision scope mounted to a twin set of Weaver bases and RPA .223 action custom rifle.*

BELOW: *The Burris laser range finding scope mounted on the Famine .308 Sako subsonic rifle marries up the bullet's trajectory downrange with the reticule.*

Chapter 5
Triggers and Safeties

Having spent a fortune on a quality barrel, stock and its action, it would be foolhardy not to consider at least the merits of some trigger work or opting for a complete custom unit. Some triggers definitely need work from the start, especially if you are converting an old military Mauser for sporting use, as this heavy, two-stage trigger has no place in the game fields. All is not lost, even with this, and a good custom gunsmith can usually lighten the trigger pull and eliminate most of the creep associated with these triggers. Even so, the triggers on factory rifles are now far better than they used to be.

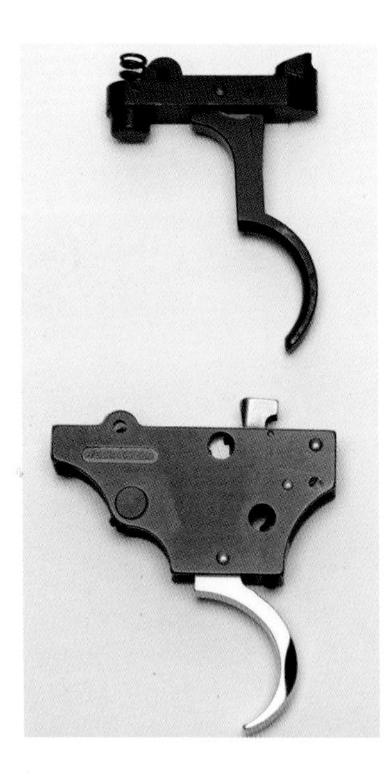

A modern fully adjustable Recknagel trigger shows how far development has progressed since its Mauser military two-stage equivalent.

Customers demand more these days and the trigger is another aspect of rifle design that has been transformed. Firms like Tikka, Sako, Anschütz and Lithgow always offered good triggers for their rifles, although some of the American firms lack a certain amount of finesse, Sako offers balanced, adjustable triggers to suit any normal hunter's needs, but even these can be 'tuned' to smoother use. With its target pedigree, Anschütz has always produced excellent triggers that do not need further customizing. In recent years, though, the triggers attached to American rifles have definitely improved. Gone are the 'litigation safe' items that plagued the industry and there are now generic factory units that offer safety and lighter-weight trigger pulls. These start-

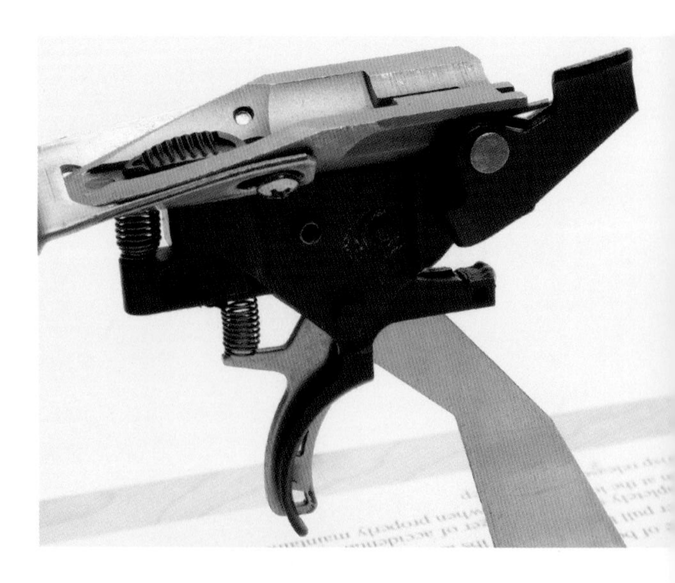

The Accu-trigger from Savage has improved the standard available in factory rifle units, but custom builders always want something better.

ed with the Savage AccuTrigger, which has an additional inset trigger blade within the main trigger that has to be depressed in unison with the whole trigger in order to fire.

However well a gunsmith tunes or hones a trigger unit, it will never be the equal of a pure custom trigger. When I first became interested in custom rifles the choice was limited to Timney and Canjar triggers from America and Recknagel from Germany. Since the explosion in popularity of benchrest and long-range shooting, however, specialized firms now offer superb trigger units to maximize accuracy, but at a price. Most common of these has to be the Jewell trigger, which is synonymous with extreme rifle accuracy and became popular through its use in benchrest quality guns in America. These are still considered to be the best triggers money can buy, but many new firms have been offering similar quality at a reduced price.

SAFETY TYPES

Wing

This term usually refers to a safety lever mounted to the bolt shroud, which is the metal covering at the back of the bolt to stop gases escaping from a pierced primer and hitting your face. They act directly onto the bolt cocking piece to move it away from the trigger sear or lock it in place. It can be designed in many ways, for example with a flip-over top design, like the older Mausers, or more commonly as a side-mounted lever. Due to the traverse range of the lever, it can also have several positions so you can select total safe where the bolt is locked down and sear disengaged, or it can be positioned at a midway point to allow the bolt to operate for unloading purposes, while keeping the trigger on safe.

When thinking about choosing a wing safety for a custom rifle it is most important to think about function as well as styling. The latter can take many forms from a bold squared look with a large articulate safety lever for fast practical use to a slimline or well-scalloped shroud with trimmed-down or ventilated levers for purely aesthetic looks. It is always best to choose one that is side operated, otherwise you will have issues with the safety lever on a flip-over type fouling the scope, if fitted. Existing wing safeties like those on the Mauser or Winchester-type actions can be resculptured or aftermarket shrouds and safeties can be retrofitted, such as those from Recknagel.

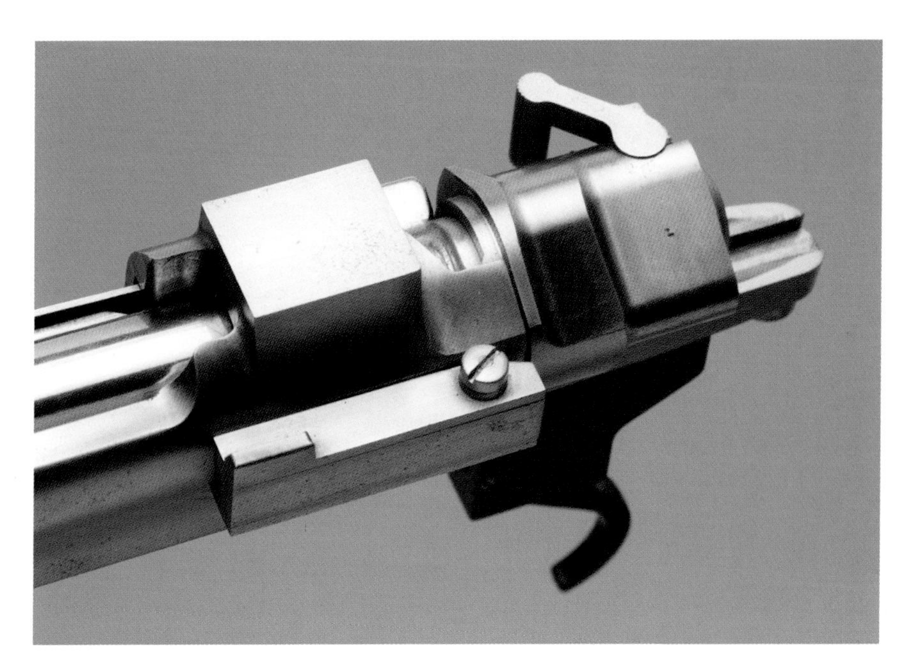

The wing type safety seen on this Mayfair custom Mauser action harks back to a traditional safety design, but with refined engagement.

Tang safeties, as fitted to the Ruger No. 1 rifles, are very convenient but some custom work is needed to avoid hang-ups with rimmed cases, such as this 450/400 Nitro Express.

Tang

Tang-mounted safeties, as on the Browning actions, need little modification and feature a very convenient position for the shooting hand to operate. Most common tang-mounted safeties are found on rifles like the single-shot Ruger No. 1, where the tang is inlet into the stock and allows a thumb-operated fast safety action. Here modifying the safety not only looks good but can benefit the operation of the rifle. Ruger safeties are usually too long for the satiety recess and also too high. This has the problem that it stops the case fully ejecting, especially if the case is rimmed, like my .450/400 Nitro Express. This is fine on the range when zeroing, but will not do when a Cape buffalo is speeding towards you and you need another shot. Reshaping, lowering and rounding off the Ruger's safety helps, especially it set off with some engraving or chequering.

Lever Safety

This is the most common type of safety you will see on a rifle. Factory rifles often have a side-lever safety as it tucks it out of the way and is still in easy reach of the trigger finger for operation, while cleaning up the appearance of the bolt and rear tang area. Most bolt-action custom rifle projects use this type and there are many aftermarket triggers that employ this method of safety. The type and style will be dictated by which trigger unit you use, but one of the most crucial feature of a custom hunting rifle is that it must be silent in operation so as not to spook the game. Keep it simple: the rifle's trigger sear is of such small proportions that there is only this piece of metal between you and the rifle being fired.

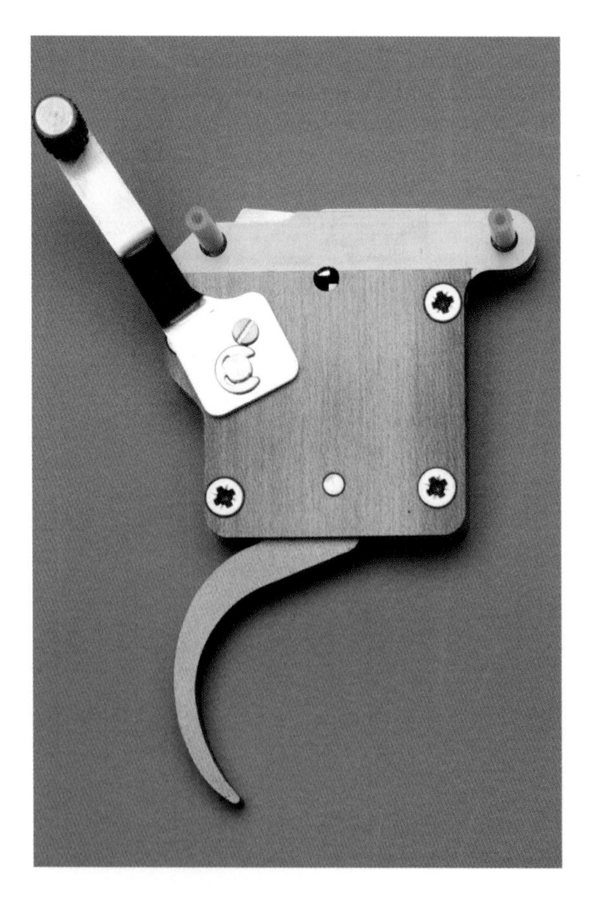

Lever-action designs, as on this Jewell trigger, are the commonest and most popular among custom rifle builders.

CHOOSING THE CORRECT WEIGHT, PULL OR TRAVEL

Weight

Dependent on the intended application, the weight of the trigger pull is largely influenced by the nature of the final target. If it is to be a long-range, tack driving varmint gun shot from a static position, then a safety is not always a consideration as you are shooting prone but the weight needed to trip the sear to fire the gun is important. Here the lightest weight is often very desirable, because as soon as a crow lands you want to send a bullet downrange before you or the rifle moves. On a normal hunting rifle, such as one for deer with a weight of about 3.5lb, this is fine for field use as you will probably be cold, wet, tired and wearing gloves. For varmints, long-range and target use a figure of 8oz is very handy as it allows a fine predetermined release with no time to pull off from your aim, which is usually best from a rifle fitted to a bipod in a static position.

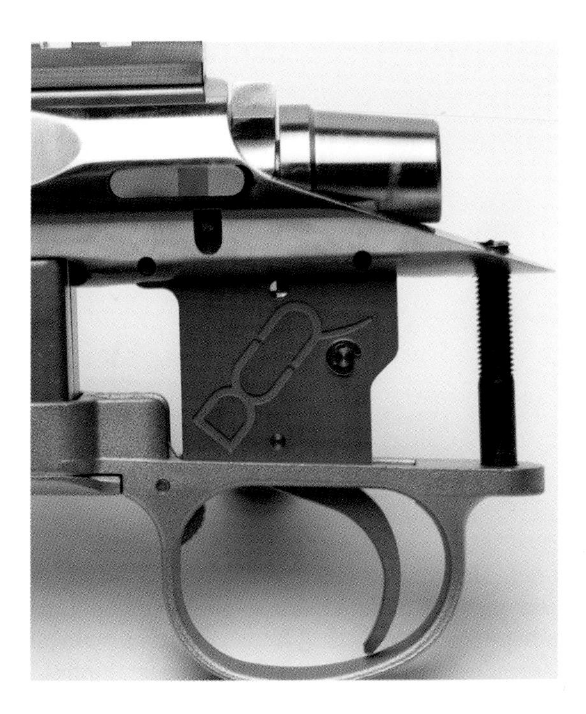

Hybrid triggers from DCR are capable of transcending the gap between benchrest and hunter trigger pulls and weight.

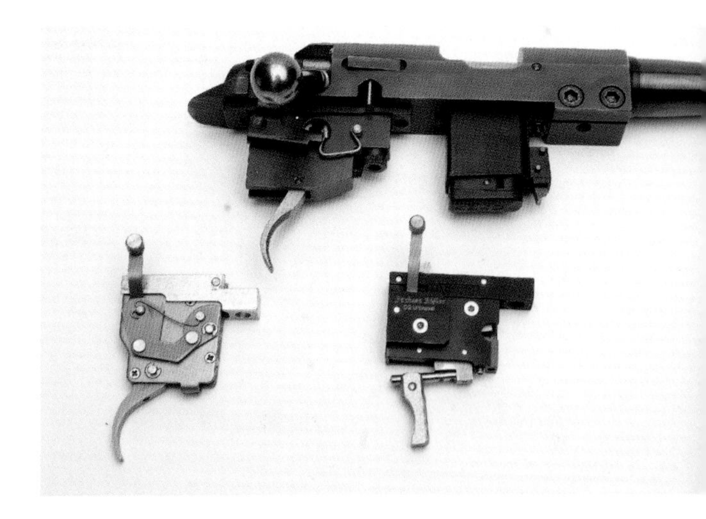

Both one- and two-stage trigger pulls have their supporters. My preference is for a single stage and straight trigger blade.

Pull

This is again a matter of preference dictated by the final scenario in which the rifle will find itself. Most military-style rifles have a two-stage trigger pull. This is because the first stage is the 'take up' or settler stage to prime yourself on target; the final or second stage is when the trigger sear trips and the rifle fires. This gives a degree of flexibility to the trigger pull whether to continue or reverse the decision to fire.

Most varmint or deer rifles – in fact most sporting arms – have a single-stage trigger, which I personally prefer. Here there is no first-stage take-up of pressure. The trigger sear trips as soon as the required weight is achieved. It realistically allows the shooter to sight, take aim and then take the shot without any slight indecision that will alter the bullet's impact. A pull of about 3lb, crisp without any drag or creep, is fine on a custom rifle used for real-world stalking exploits. People, though, always want better and assume a lighter trigger pull is best. I would probably agree with that, but unless you know how to use it you will find that premature ignition is a problem. I would prefer to fit a quality trigger that is adjustable within the pull weights you need and set to a level that suits your style of shooting. Target shooting from a bench, while developing an accurate load, is quite unlike what you will experience when out hunting with freezing fingers.

Travel

The travel of a trigger assembly tends to be forgetten, but it can be an important part of the whole process of pulling the trigger. Follow-through is the precise and consistent nature of pulling the trigger. It should not be a snatch or jerk but a considered, slow and deliberate pull-through of all the stages of the trigger cycle until the trigger blade comes to a rest. Here the travel is important. A continuous travel can often allow you to 'over-pull' the trigger and jerk the shot, while a clean stop allows just the sear break and no more travel. This can make a big difference as it trains the brain to act one way, and one way only, to achieve a consistent and smooth trigger release and ultimately better accuracy.

TRIGGER GAUGES

These tools are invaluable for checking the correct trigger pull on a custom trigger after adjustment, and also to check that the adjustment is consistent after a period of shooting. They can take the form of either mechanical or electronic read-outs. Both types use a steel trigger probe that acts like your trigger finger to engage the trigger blade, and then a hand-held read-out is pulled rearward to indicate the trigger pressure. A mechanical unit has a scale on the outside of the handle and an indicator attached to a spring that, under tension from the trigger, indicates the weight when the sear breaks. An electronic digital unit has the advantage of an easy-to-read screen that uses strain gauge test technology to produce accurate pull weight measurements from 0–12lb weight. It also has the advantage that it can average ten readings for a more consistent overall trigger pull from your rifle.

An over-travel screw, as on this Contender G2 rifle, stops the trigger blade moving too far rearward and defines a pull and trigger travel.

Trigger gauges are essential when adjusting triggers to a certain pull weight, but I prefer to adjust to what feels good for me.

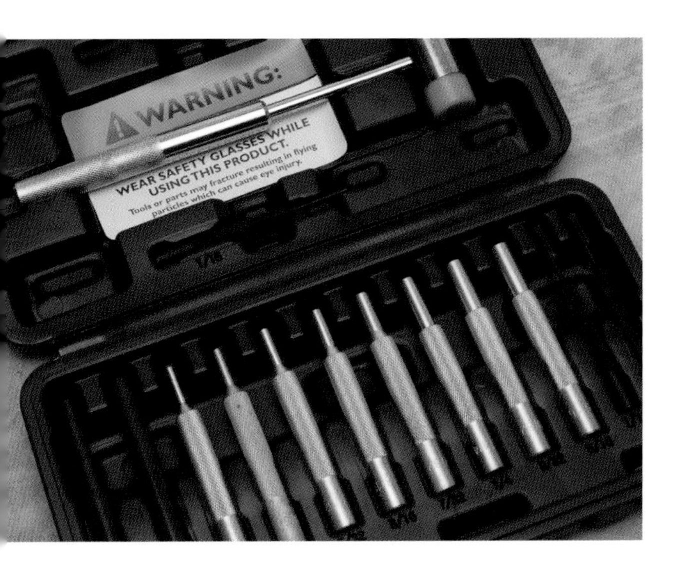

A set of good punches, while very handy for most custom rifle work, is essential for trigger replacements.

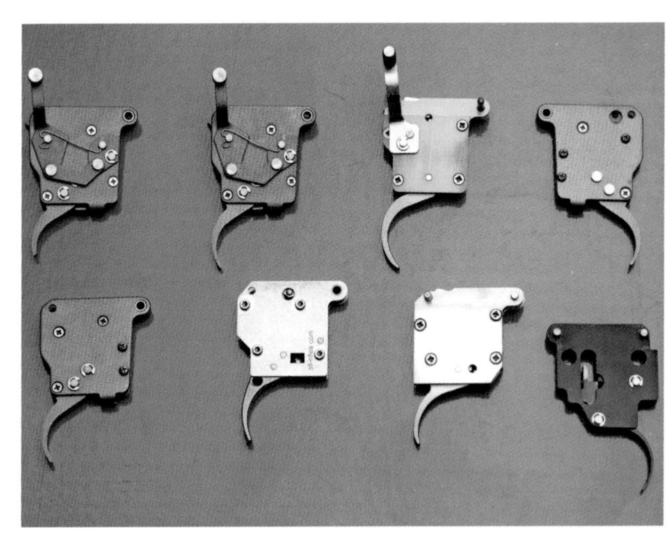

Custom trigger units are expensive but they transform a rifle instantly to provide an instant and safe sear release which ultimately results in improved accuracy.

PUNCHES

Invest in a decent set of punches to remove the securing pins commonly found in trigger units and many other firearms and air rifle parts. A good set, which will set you back about £40, is made from hardened tool steel for long life and ground to give a 5 thou clearance on the pin hole; a set should cover the common 1/16in, 3/32in, 1/8in, 5/32in, 3/16in, 7/32in, ¼in and 5/16in diameters needed.

LUBE

Don't lubricate the trigger. Stainless steel triggers are meant to run dry: any lube can affect performance as it allows debris, fluff and dirt to accumulate and cause a malfunction.

CUSTOM TRIGGER UNITS

This is where I believe you should be spending your money as the trigger is your only real contact onto the firing mechanism. It is crucial that the point that you wish to release the trigger is actually the time the sear drops and the sequence is initiated, otherwise accuracy and consistency will always be compromised. It is possible to hone a factory trigger and a really good gunsmith can work miracles on it, but you will find that, after going through the process of having the action trued, you might as well have bought a custom trigger from the start.

A good custom trigger will cost between £100 and £250. It will instantly transform a custom rifle and add to its resale value. Manufacturers offer triggers of the same basic design but with differing features to suit benchrest, long-range or hunting rifles. Each style has its individual weight and pulls, and comes with or without safety hook-ups. Due to the many types of rifles in use, it is impossible for custom trigger makers to devise one that will fit all the varying actions or 'hook-ups'. That is why most custom actions have a universal hook-up set for the profile of a Remington Model 700, as this remains the most common action for customizing. As many actions are Remington 700 clones, it makes sense to have the same profile machined to the underside of the action to fit the custom trigger too.

Timney triggers have been available for a while, but these Riflecraft-supplied models for a Remington and Winchester are well priced and offer good performance.

Replacing a standard military trigger with a Timney single-stage unit transformed the accuracy of this K98 rifle.

CUSTOM TRIGGER MAKERS

Timney

Allen Timney established Timney Triggers in 1946 at a time when many shooters were returning after the war and were converting military rifles for sporting use. Better triggers were needed as military double-stage units were woeful for this purpose. Timney's first triggers were for Mausers, Springfields and Enfields, which were the mainstay of the military forces in the Second World War, but the range was expanded as his business grew: the company, based in Arizona, now offers more than 170 different trigger models for rifles, shotguns, semi-automatic weapons and even triggers for archery.

The typical non-safety trigger for an old Mauser K98 action illustrated here can be used to customize an old service rifle to vastly improve the trigger pull. When converting a Mauser action to sporting use this Timney trigger will transform the rifle's performance. When fitted to a BNZ K98 sniper rifle built in 1944, for example, it improved the feel and performance of the rifle to shoot 1.5in groups at 100 yards with reloads.

CD Universal

This quality trigger was devised by the French rifle designer Robert Chombart and has been updated by Peter Jackson from Jackson Rifles. I have been really impressed by the quality of manufacture and performance, which offers far better and more refined trigger pull, weight and travel attributes than any factory unit. They are not cheap, but in my view are worth it. Key

The CD Universal trigger, available from Jackson Rifles, vastly improves both full-bore and or small-bore rifle trigger pulls.

features are that every part has a hard anodized aluminium housing with stainless steel fixtures and ball bearings at all key friction points. The D-2 sear levers are tool steel and heat treated for longevity. The trigger blade can be ordered with a gold anodized trigger shoe in either a deeply curved type or as a straight model, which can be reversed to present a slightly curved surface. My trigger has the straight version and it was perfect for use on my Sako Finnfire custom rimfire rifle.

Recknagel

The German company Recknagel produces all manner of custom parts to the highest standards, including premium quality triggers for Mauser- type hook-ups. These are the first choice for custom gunsmiths wanting to replace the old factory or military trigger on Mauser 98 actions. The range of triggers is staggering as are the differing finishes, sizes of trigger blades and even their profile.

A Recknagel steel-bodied, slim trigger blade fitted to the precision-built Mayfair action, a match made in heaven.

The German company Recknagel is synonymous with supplying precision aftermarket triggers for Mauser actions. Here is the single-stage model with 6mm profiled, gold-plated trigger.

The model 30010-1300 is a K98 trigger that combines a standard single-stage pull with that of a single-set option. As a single stage it breaks incredibly smoothly at 1400 grams weight and in the set position it is light at 300–400 grams. Both are adjustable, with heat-treated internals for safety and excellent wear characteristics. The trigger is available with a steel or aluminium body. This model also has a gold-plated trigger blade of 6mm width and lightly profiled shape, ideal for any custom Mauser rifle. Other models with side safeties are available, again in many styles and trigger options.

Jewell

These are widely considered the crème de la crème of rifle triggers and adorn many a custom rifle used by all types of shooters. They are designed to be fully adjustable and provide a light, reliable trigger pull for extreme accuracy. The self-contained trigger unit replaces the original

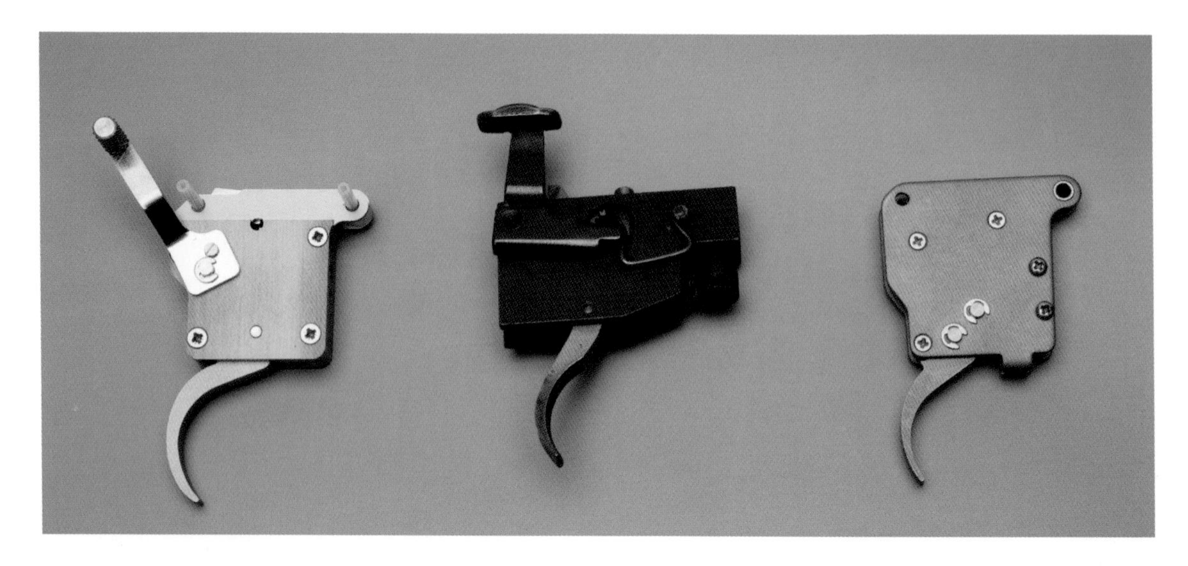

Jewell triggers are the standard by which all others are judged.Varying trigger weights can be chosen to complement benchrest, hunting or varminting scenarios.

factory unit. Jewell triggers share a common chassis but utilize a different hook-up adapter to fit varying action bottoms. All parts are CNC machined for close tolerances and quality. The housings are 300 series stainless steel with all the internal parts made from 440 stainless steel that has been hardened to 58 RC for longevity.

The HVR model has a 1.5–48oz trigger pull weight, while the BR version fits right- and left-hand rifles and has a very sensitive 1.5–3oz weight for extreme precision. The RHVR or Remington Hunter Varmint Rifle trigger is designed to fit Remington 700 variants (custom action clones or 40X actions) with a top lever safety and bottom bolt release. Common to all is the single-stage release design, which has a three-way adjustment to set pull weight, let-off and over travel.

RPA Quadlite

I use two RPA Quadlite actions, one with a .223 bolt head and one with a .308 bolt head for testing new wildcat cartridges. These both have Quadlite trigger units fitted and are a two-stage design as required by target shooters and preferred by the military. The operation is silky smooth with a predictably consistent release point. They are available in 0.5kg and 1.5kg pull weights. All models

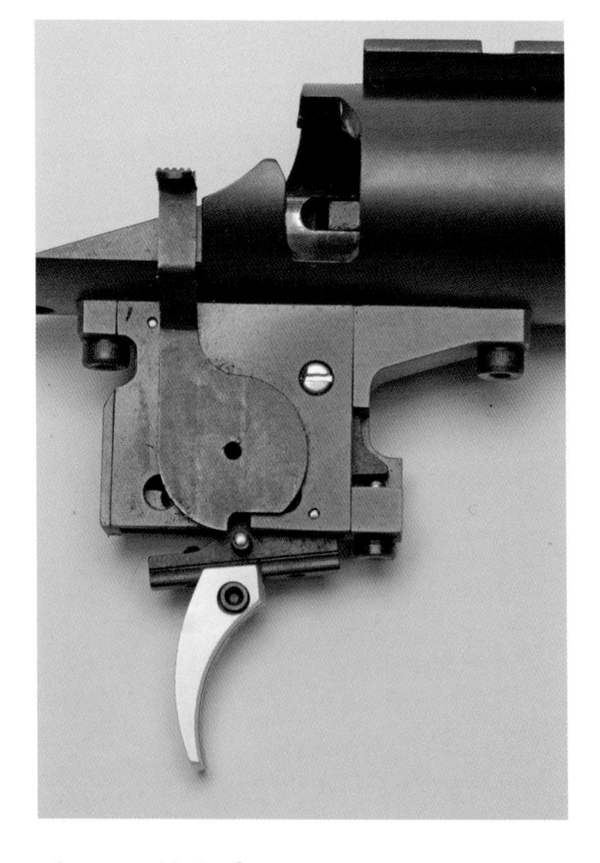

The trigger blade of RPA triggers can be adjusted for cant and length, which really enhances shooter compatibility.

can be fitted with a top-mounted safety and have an adjustable finger pull lever. Early versions of the RPA Quadlite trigger had a bottom-mounted safety, but I do not like this as it is positioned too close to the trigger blade.

Cadex

This is another replacement Remington trigger that offers near-benchrest quality at a good price. There are two models available. The DX1 single-stage trigger has an adjustable pull weight and travel. The trigger shoe is also adjustable so you can lengthen or orientate it to suit your trigger finger's position. The aluminum design is light and precision made. Multifaceted openings to the trigger housing facilitate the removal of dirt or water that could cause a trigger malfunction in the field. The DX2 double-stage trigger has an adjustable pull weight in both first and second stages, as well as an adjustable second-stage travel. These very well-made triggers offer an alternative to the growing number of aftermarket Remington replacement triggers.

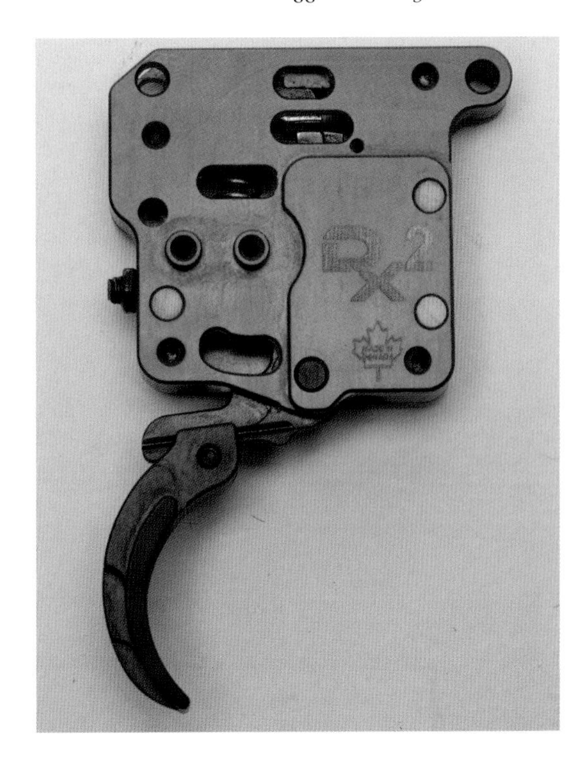

Cadex is a newly designed trigger mechanism that offers a very precise trigger pull and let-off.

Trigger Tech

Trigger Tech's design presents a radical change from normal trigger technology. Most triggers have some form of creep or friction problems on their sliding surfaces, which are usually customized by altering pivot points and polishing alternative sear types. Trigger Tech, however, approaches this differently by fitting a patented free-floating roller between the sear and the

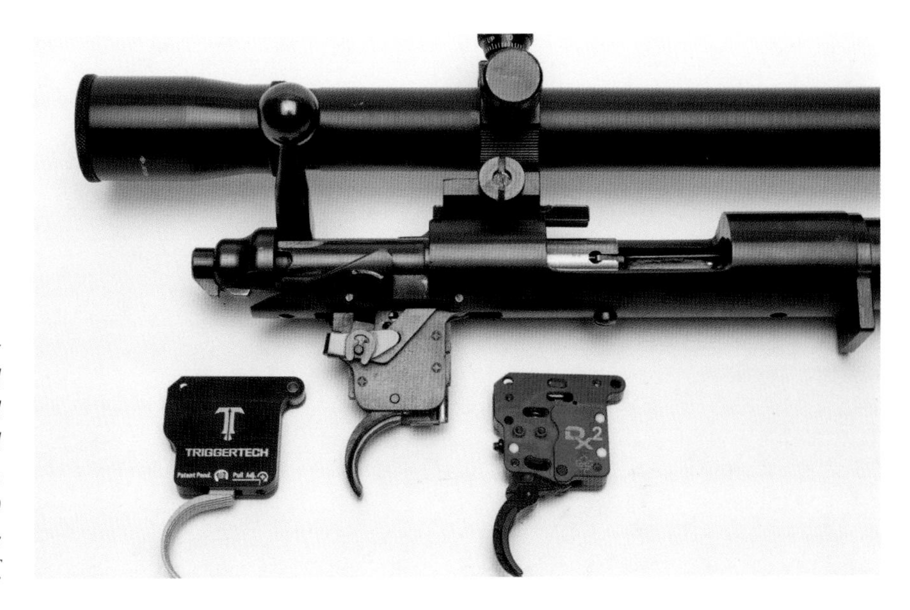

Trigger Tech is the new Timney, affordable and with very good performance. I fitted one to a Remington 40X .22LR for 200 yard use and groups shrank by half.

trigger engagements. The firm calls this Frictionless Release Technology. By eliminating the sliding friction points the trigger release is now smooth, consistent and safe. The trigger pull feels lighter and drag or creep is eliminated. This has the effect of transforming the trigger operation so that the bullet is fired without delay, potentially increasing the lock time between thought and action and so enhancing the accuracy. Wear between the crucial surfaces of the trigger components is also reduced and that enhances safety because they last longer. Other than the frictionless roller, the main parts of the trigger are made from surgical quality stainless 440C tool steel on the key sear, latch, trigger and safety parts. I have one fitted to a Remington 40X .22 rimfire that I use for long-range, 200-yard rimfire silhouette shooting. It has transformed the accuracy as the lightest of releases is needed to fire the rifle and so there is no time for me to alter the aim.

Rowan Engineering

Rowan Engineering Limited specializes in precision-made custom parts and accessories for air rifles. It also provides CNC machining services to industry, automotive sectors and research and development organizations. Its components include trigger blades, guards, silencer adapters and recoil pads. Brass trigger guards and aluminium polished or satin-finished guards really enhance the appearance of a standard air rifle.

I really like the trigger blade replacement units. I have used them on a couple of Venom custom air rifles and found that they greatly enhance the trigger pull. They are easy to swap over for the factory unit and are available in many guises. The set-back trigger look, for example, is aesthetically appealing, moving the trigger blade further back towards the pistol grip. On the standard unit the original safety ad-

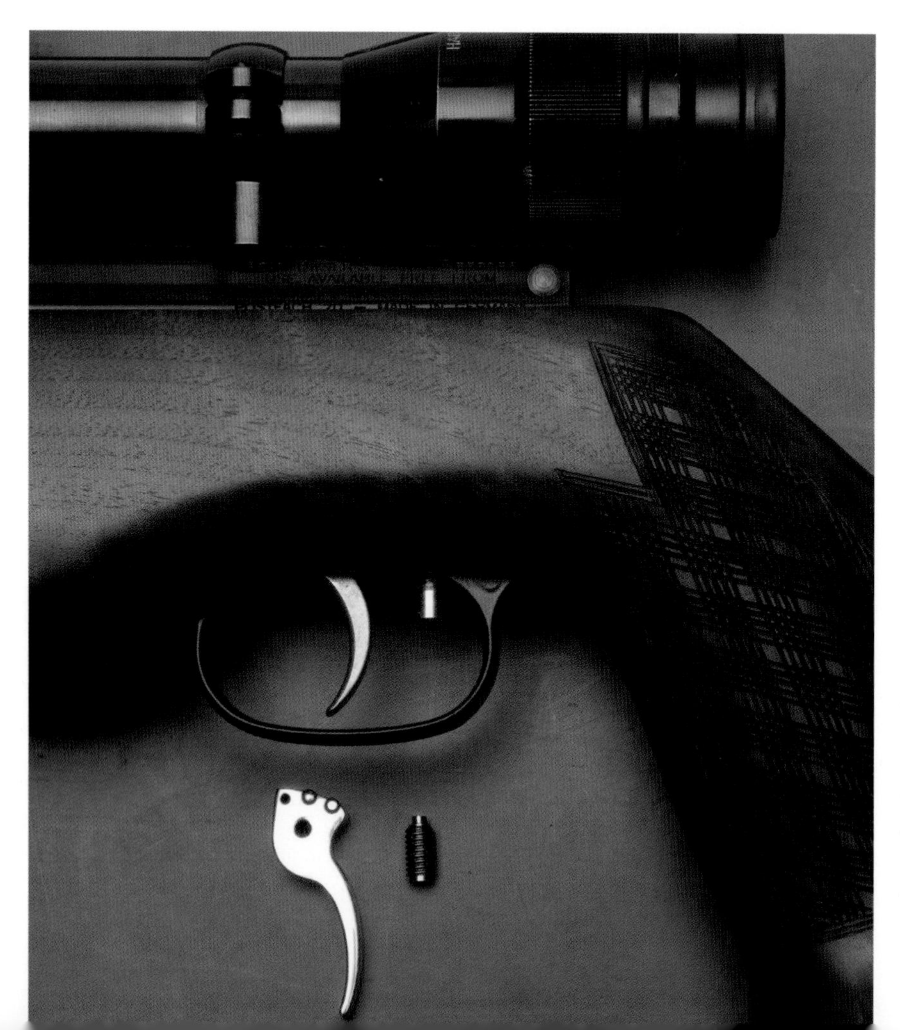

Rowan Engineering, based in Banbury, offers custom parts for air rifles. Their brass, aluminium or black set-back triggers transform the look and feel of any custom air rifle.

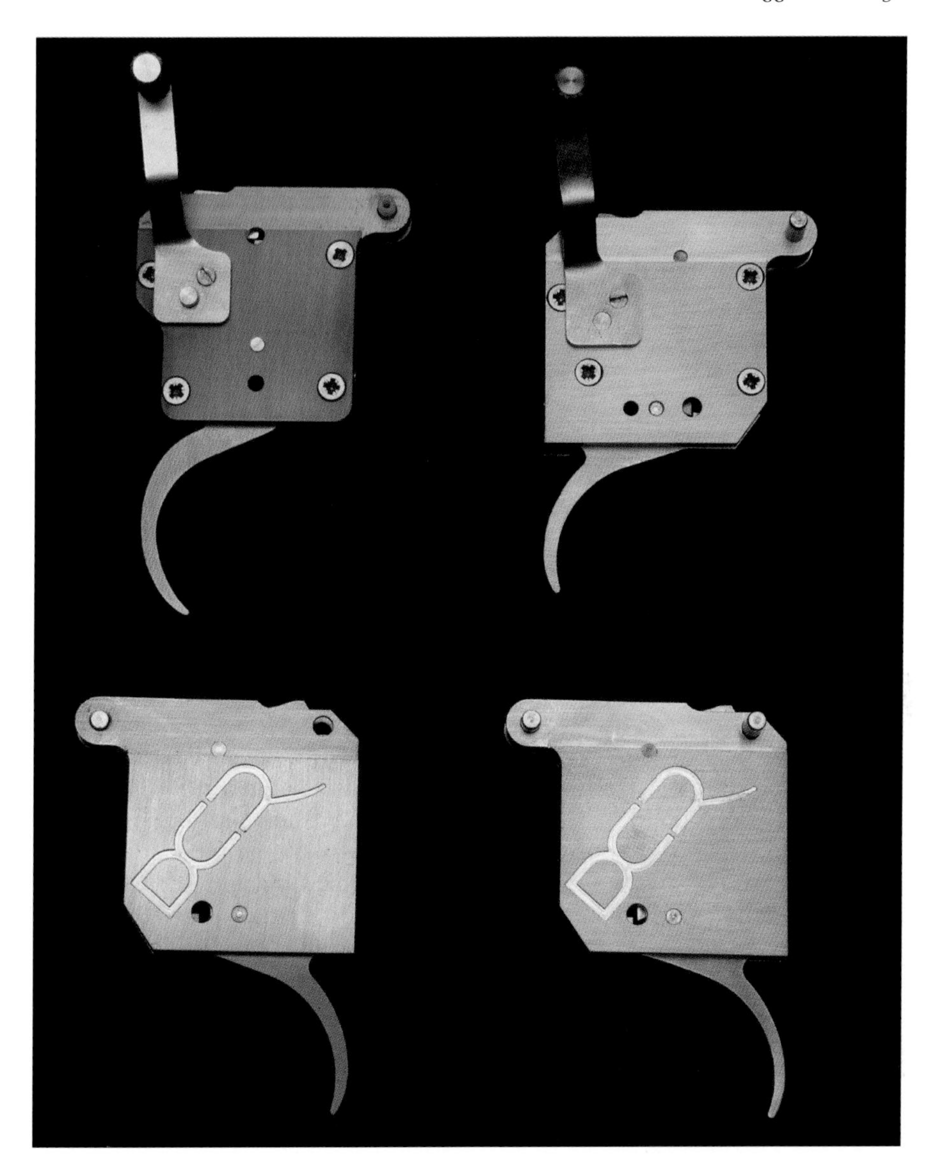

DCR makes a comprehensive range of triggers for custom rifles, ranging from hunter class to pure benchrest weights and the new hybrid for a cross-over target/hunter trigger.

justment screw would get in the way, so Rowan provides a shorter screw to allow for this. The replacement units are available with a curved or flat blade, and also in brass or aluminium. They can make a normally bland part of a custom air rifle look and function much better.

DCR

Devon Custom Rifles makes a range of precision triggers for all forms of shooting disciplines. There are three current models. That intended for the benchrest target disciplines has 2oz to 6oz let-off weights for a precise and instant trigger sear release. A standard trigger can be set from 1.5–3.5lb weight and a new hybrid precision trigger covers from 6oz to 1.5lb weight. These cover every possibility and provide a home-grown alternative to triggers from abroad.

Chapter 6
Feeding a Custom Rifle

After taking months or years to create your custom rifle, it would be pointless to shoot just any old ammunition and fall at the last hurdle. Using inferior ammunition will undo all the custom gunsmith's work and your dream rifle will soon turn into a nightmare. Having said that, factory ammunition has increased in quality to meet shooters' expectations, drawing on new manufacturing techniques combined with advances in powder and bullet performance. I have tested many rifles where factory ammunition has outshot reloads, which is a bit embarrassing but shows how, by choosing carefully, your new rifle can be fed a good diet if you do not want to reload.

If you have gone the distance, however, and spent your hard-earned money on a custom rifle, it would be to your advantage to invest in precise reloading equipment and learn some techniques that will glean every last inch of accuracy and longevity from your rifle. It need not be difficult. The basic reloading techniques may be found in my earlier book *Sporting Rifles* (Crowood Press, 2009), but here are a few pointers to the advanced equipment and procedures needed for extreme accuracy.

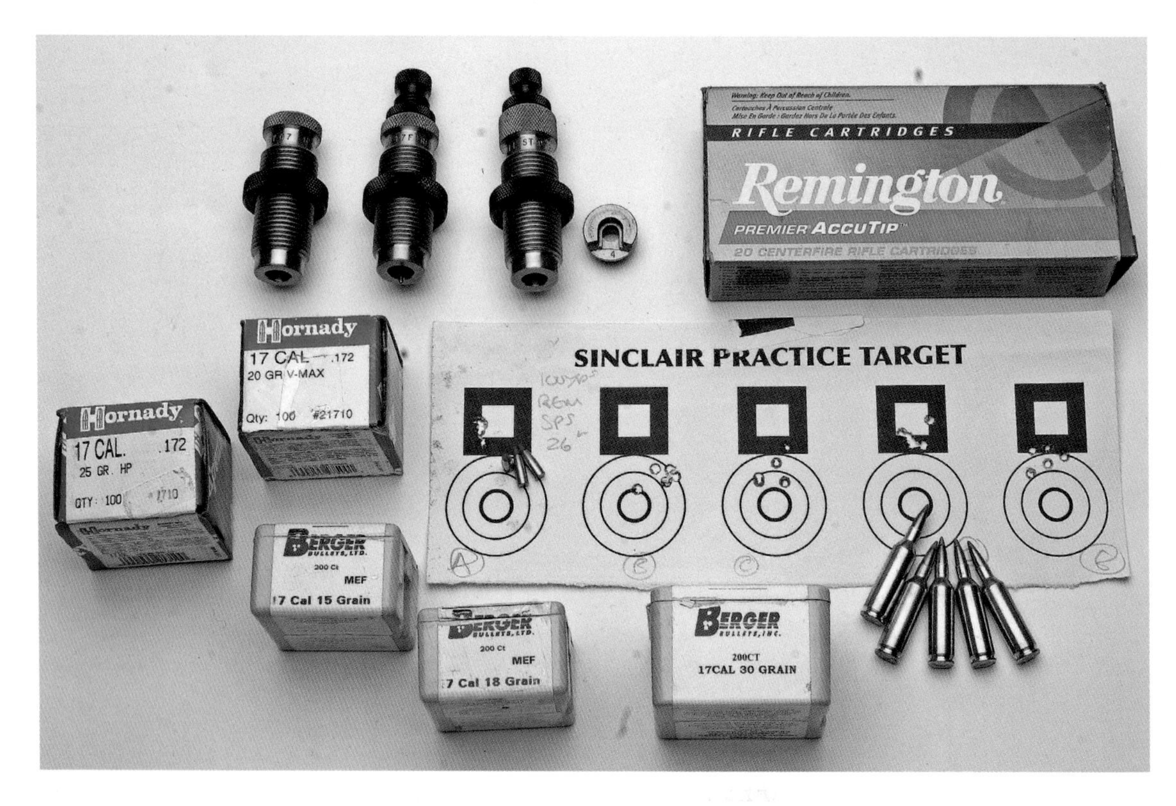

Getting your rifle to shoot correctly is now paramount, otherwise all that hard work has been wasted.

FACTORY AMMUNITION

Centrefire

Although reloads will always be a better means of achieving optimal accuracy in your new rifle, there is nothing wrong with using factory ammunition. It largely depends on the client and their intended use of the rifle. For those who use their rifle for stalking only once a year it makes sense to use only factory ammunition. Similarly, if you intend travelling abroad factory ammunition is often easier to pack and get through customs, whereas reloaded wildcat cartridges can cause problems. The general rule when choosing any factory ammunition is to match the bullet weight, style and profile to the custom rifle's barrel twist, chamber dimensions and end use. There is no point having a slow-twist custom .308 for lighter, fast-expanding bullets and then shoot heavyweight 180 grain bullets through it. It is always best to choose a popular and readily available brand that is of current production, not old stock.

Ammunition is usually supplied with twenty rounds per box. Buy at least five boxes from different makes with the bullet weight and style you need and see which shoots the tightest groups. The rifle will tell you what it likes, not the salesperson. With this preferred load, go back to the gun shop and buy as much of that ammunition as you need, making sure you get exactly the same lot number, which can usually be found typed on the inside of the folding closure.

A more adventurous course of action is to take advantage of your new rifle's specific chamber and throating length (leade). Factory ammunition is made to standard specifications that will fit the average rifle's chamber for a particular calibre, such as those specified by the Sporting Arms and Ammunition Manufacturers' Institute (SAAMI). It is also usually of one uniform length, dependent on bullet weight, that will fit and function through a rifle's magazine. That means you are shooting average-length cartridges through a custom chamber that will certainly place the bullet too far away from the rifling lands and therefore reduce the accuracy potential. Many custom rifle builders understand that their clients do not always want to reload, so they offer a custom reloaded ammunition to suit the individual's rifle.

Even if you do not reload, there is still a good range of factory-loaded ammunition to choose from. Find a load that shoots well and then buy more of the same lot number.

Ballistic programs like QuickLOAD and QuickTARGET have revolutionized the accurate predicted reloading procedure. When you buy a new rifle you can order a quantity of readymade ammunition to complement it. This

Factory ammunition can be adjusted to suit your rifle's chamber with a bullet puller and seater die.

is a good way to go as any enterprising custom gunsmith will always work up a load development plan and most accurate load for your rifle as a matter of course. They will charge for this load development, but it is worth the cost to save the time it takes to find the best load.

One easy technique to improve factory ammunition is to use a kinetic bullet puller, although you will also need a seating die to match the cartridge, overall length (OAL) gauge and reloading press:

- Insert the OAL gauge into the chamber and accurately measure the overall length from the case base to the ogive point on the bullet, that is where the bullet contacts the rifling.
- Measure this distance accurately with a set of digital callipers.
- Place your factory cartridges into a kinetic bullet puller and tap briskly as per instructions on a hard surface. This should move the bullet out of the case a small amount. Repeat until this length is more than the rifling contact distance measured earlier.
- Now place this lengthened case into the seater die and reseat the bullet to exactly the same measurement as it contacts the rifling.
- From this point onwards you can reduce the OAL to increase the jump from bullet to rifling, for example at intervals of 5, 10, 20, or 30 thou off the lands.
- Shoot groups for each new OAL to tune your new factory load for the best accuracy and function in your particular rifle.

One word of warning, however, is that some bullets, including Barnes all copper, GMX and E-Tips, shoot better at 50 to 70 thou off the rifling lands, so take care.

Rimfire

Do not forget rimfire rifles as many custom rifles are based on rimfire cartridges such as the .22lr and the newer .17 HMR. Even though these rounds are not reloadable, you can still maximize accuracy by a few simple steps.

Sorting by weight

This is an easy process that can be described using the example of a CZ 455 .17 HMR rifle. This shoots sixteen shots for best accuracy, so I weighed the fifty rounds in a box and took the sixteen that were most closely matched in weight for the test. I used an Ohaus GT480 scale and I repeated this twice so I had eight groups to compare. The average weight of fifty rounds was 2.601grams, of which sixteen had

Weight sorting 17 HMR factory ammunition can maximize the accuracy potential.

a uniform weight of between 2.598 and 2.604 grams. The highest figure was 2.611 grams and the lowest was 2.592 grams, so pretty uniform really, with no outlier results. These were shot as four groups of four shots, which averaged 0.40in spread at 100 yards. These were then compared with the sixteen rounds that were shot in the same way but were picked at random from the box of fifty rounds. These averaged 0.55in groups, so there was a significant increase in accuracy. Velocity too was harmonized with a standard deviation of only 7 for the sorted batch or 16 for the unsorted batch (the smaller the figure, the more uniform the results). This was repeated with a different batch and very similar results were obtained. This may seem a little long-winded, but it is a cheap way of improving accuracy in your rifle.

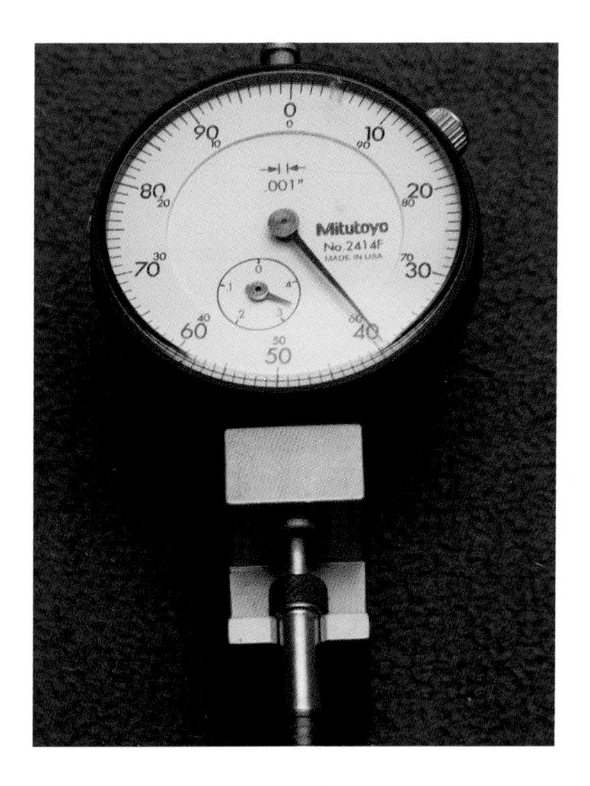

Rim thickness gauges are used to sort out inconsistent rims.

Rim thickness gauge

Rim thickness sorting on rimfire ammunition can also be beneficial. I use a Bald Eagle rimfire gauge for .22lr ammunition or with a special .17 HMR adapter that is available to sort the ammu-

Batch 1 (Average)	2630fps	261.2ft/lb	0.45in at 100 yards
Batch 2 (Average)	2624fps	259.9ft/lb	0.40in at 100 yards
Batch 3 (Average)	2674fps	269.9ft/lb	0.35in at 100 yards
Batch 4 (Average)	2607fps	256.6ft/lb	0.42in at 100 yards
Batch 5 (Average)	2612fps	257.6ft/lb	0.35in at 100 yards

nition into consistent rim thickness. Using the same procedure as the weight sorting described above, the best sixteen from a fifty-round box were compared to the unsorted batch and then the test was repeated to confirm the results. These were less spectacular, but there was a 5.0 per cent increase in accuracy, down from 0.55in to just over 0.475in, and the standard deviation had dropped from 16 for the unsorted batch to 9 for the rim-sorted batch.

Sorting by lot number

Different batches or lots of the same ammunition can affect accuracy, velocity and energy figures. You can test your new custom rifle by totally cleaning the bore each time, then shoot one fouling shot followed by a string of five groups of four shots. Repeat this for five different batches of Hornady.17HMR ammunition. This can also apply to centrefire ammunition.

The table at the top of the page shows the results for the five batches of Hornady that were tested, twenty shots per batch:

There was not much difference but this shows that, if you have to rely on a factory's quality control, the three practices described above can help to achieve some form of segregation into uniform rounds that should improve on accuracy if all the components are of similar weight.

Air Rifle Pellets

Lube

You only have to use a tin of pellets for a short period with the tin lid off and soon the lead surface of the pellet becomes oxidized. This layer creates a coarse surface to the pellet, making its passage down the barrel less than smooth.

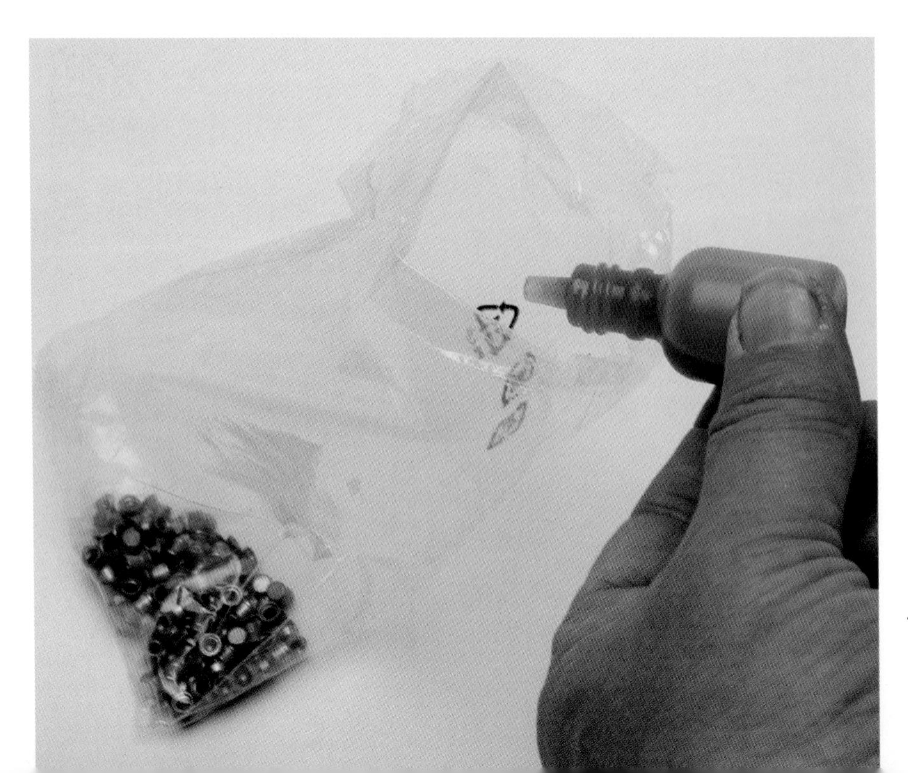

Airgun pellets can be pre-lubricated for more consistent velocity and accuracy.

A very thin layer of lube, such as Napier Power Pellet lube, Daystate Rangefinder lube or SPL (Special Purpose Lubricant), which is sourced from the US and mixed in equal parts with water, gives a uniform surface to ensure a reliable and repetitive pellet movement through the bore and improve accuracy. Claims of huge percentage accuracy increases have been made, but my experience is that it makes a good pellet that is accurate just that bit more consistent. Continued use will eventually condition the bore, like moly on centrefire ammunition, and this in turn reduces fouling in the barrel for a consistently smoother bore surface.

Weighing

Weighing pellets definitely helps. It is tedious, but there can be a 5 per cent variation in pellet weight in a tin of 500 pellets, which will certainly affect the accuracy. As with centrefire brass cases, I would segregate them into groups of 0.1 grain differences, discarding those that fall within the outer 10–25 per cent by weight (or use them only for sighting), and select the pellets closest to the mean value for ultimate accuracy.

Sizing dies

Due to the differing internal bore sizes of the barrels available to air rifle shooters, such as 5.5mm or 5.6mm in .22 calibre, most airgun pellets makers have varying sizes of the same pellets to fit different barrels. Even within the same barrel dimension – for example .177 calibre makers offer 4.50, 4.51 and 4.52mm sizes – you can fine tune a pellet to your own barrel, rather like reloading for a centrefire rifle. You can, however, make them even more uniform by sizing each individual pellet in a calibre and size-specific sizer die. Here a pellet is swaged through the die so as not to deform it but yield a uniform and true pellet specific to the diameter you want. This way you now know that each pellet is identical to the other. This is very handy for .20 calibre pellets where expanding the skirt can be beneficial, as with the Venom/V-Mach .20 die or a T. Robbs .25 calibre pellet swager to make uniform pellets in this large calibre, since the skirts are often dented or deformed.

Airgun pellets are often not the right size for specific barrels, which makes this V-Mach pellet sizer or expander in .20 cal particularly useful.

BULLET CONCENTRICITY GAUGE

This is a quick fix for eliminating odd ammunition, both factory sourced and reloads, that is out of true. Factory ammunition is loaded on an assembly line and sometimes the bullet can be seated on a cant due to irregularities in the bullet, loading dies or variations in the neck thickness of the case. This may be only a small amount, but sometimes it is enough to start the bullet's journey down the barrel at an angle. There are many gauges available to check if the bullet is running out of true. If it is you can reject that cartridge and on some gauges you can correct the tilt to true up the bullet alignment. This type of gauge also checks neck run-out and concentricity.

Sinclair Concentricity Tool

The Sinclair concentricity gauge is very versatile and can measure cases all the way up to 50 BMG. It is very well made and measures run-out by rotating the case or loaded round by resting it on two sets of bearings that are set in anodized aluminium blocks. These bearing blocks can be adjusted for differing cartridge lengths. The bearing blocks sit in a milled slot in the anodized base plate, which provides a stable platform allowing accurate readings. The milled slot keeps the blocks in alignment with each other, which in turn keeps the cartridge well supported so you can spin the case smoothly for an accurate reading. There are large thumb levers on the bearing blocks, making adjustments quick and easy. You can measure sized cases on the neck and loaded rounds on the neck or run out on the bullet. An indicator tower gives precise measurements for vertical and horizontal adjustment of the dial indicator. The mounting block for the dial indicator is designed to accommodate dial indicators with standard 0.375in mounts. This is great for factory ammunition too, but I use it for checking if I have buckled a case neck when reloading a round, which may not be obvious from a visual inspection. It is also essential to check the neck especially after necking up or down in the course of making a wildcat round.

The Sinclair concentricity gauge measures run-out by rotating the case or loaded round by resting it on two sets of bearings.

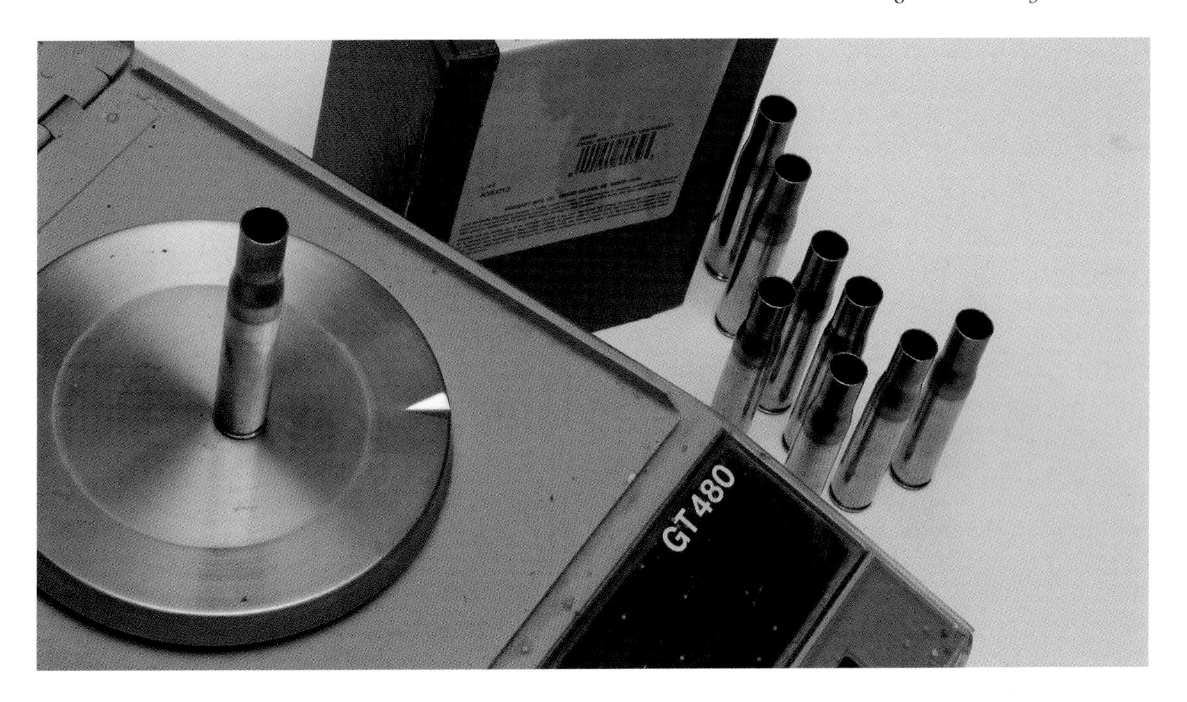

Segregating brass cases by weighing eliminates any odd cases from a batch.

RELOADING CENTREFIRE CASES

Weighing Brass Cases

This topic was covered in detail in *Sporting Rifles* (see above), but a brief account may be of help. If you buy 100 new brass cases for reloading, make sure their individual weight is as close together as possible. If they are not it indicates that their dimensions, in particular the internal dimensions, will be different and this will affect the burn rate and efficient combustion of the powder. This in turn will cause the velocity and pressure to vary, having an adverse effect on the accuracy. Weigh each case and segregate them, as described above when discussing airgun pellets, and select those with the weights closest together.

Most case necks have dings in them, so a neck deburrer is a very handy tool.

Tools to Improve Cases

Neck de-burrer and reamer

A simple set of tools is used to help ease the loading of bullets into the case. Most new cases will have dents, dings, ovals and burrs at the mouth of the case neck. As a matter of course the mouth should be chamfered with a de-burring tool. Most of these are double-ended so they can be employed for internal and external de-burring. I use hand de-burrers, since I like to feel with my hands how much brass is being removed, but many can be used in a case trimmer, purpose-built case preparation station or a powered screwdriver.

Most inside-outside de-burring tools are suitable for all cartridge cases from .17 to .45 calibres; a de-burrer for .50 cal is an oversized special tool. They are simple to use. Insert the pointed end to chamfer the inside of the case mouth using its precision-ground and hardened cutting edges. I usually turn them two to three times to uniform the mouth, then flip the de-burrer over so that the open end now sits over the case mouth. Turn it a couple more times to chamfer the outside edge of the case. The mouth is now uniform and can accept bullets without the risk of buckling the thin brass neck with flat-based bullets. There are de-burrers with differing profiles for specific bullet types: VLD (Very Low Drag) bullets, for example, fare better with a 28-degree de-burrer angle, whereas the standard angle de-burrers are 45 degrees.

Firing pin hole de-burrer

A flash hole de-burring tool is designed to be inserted through the case mouth and uses the web of the case or case mouth to locate the cutter into the primer hole. The tool has a hardened cutting edge and a few turns produce a small chamfer on the inside edge of the flash hole, removing any burrs when the primer hole was made. Most are punched through and thus have burrs. Cases with cut holes fare better but they can still benefit from uniformity to their edge. The rationale behind this procedure is to ensure the ignition force radiates and spreads as evenly as possible, igniting the powder charge at the same time. Flash hole de-burring is performed only once and is most effective on brand new cases.

Primer pocket uniformer

This is a small but very important tool for the accuracy-conscious custom rifle shooter. Most uniformers are precision-ground carbide tools with at least five cutting flutes for smooth, flat uniforming of the bottom of the primer pocket. A shoulder ground into the carbide tool bit butts up against the primer pocket to manage a uniform depth to SAAMI specifications. They come in various sizes for small and large rifle primers. The aim is to make sure the pocket in which the primer sits is perfectly square without rounded edges to the bottom faces, otherwise the primer will not seat to the correct depth and result in irregular ignition. The problem is most noticeable when a case is stood up on a flat surface and it rocks as the primer is sticking out. When cham-

Brass burrs left over from manufacturing around the primer flash hole, just as it enters the powder column, can be removed with these tools from Sinclair and K&M.

It is crucial for proper ignition that the primer is seated squarely in its pocket, so a pocket uniformer tool, such as this one by K&M, is essential.

bered a protruding primer causes the head of the case to sit deeper; on ignition the case backs out of the chamber to fill the bolt face, which stretches and thins the cases.

Case trimmer

Cases must be kept to a uniform length otherwise they will become too long in the rifle's chamber. Firing elongates the brass case. If it is not kept to a specific length it can cause feeding problems, headspace issues and high pressures due to the case mouth being crushed into the bullet and not releasing it. Two makes of case trimmers I use are by Wilson and Hornady.

The L.E. Wilson stainless steel case trimmer is one of the most accurate available and my favourite case trimming tool. The case is held so that it is aligned square to the cutter during trimming, ensuring that the trimmed length is identical for each case. The case neck is held by the body taper using special case holders and the case trimming operation does not require pilots. The trimmer uses a rail system on which sit the case holders, cutter housing and adjustment stop sit, held in perfect alignment with each other. Trimmer case holders are available separately for different cartridge sizes. A mounting stand, which can be C-clamped or permanently mounted to a bench top, makes case trimming much quicker and more comfortable. Trim adjustments can easily be made in 0.001in increments.

This Wilson case cutter and the associated cartridge holders are used to deal precisely with cases that stretch on firing or if you need to remove brass on a wildcat project.

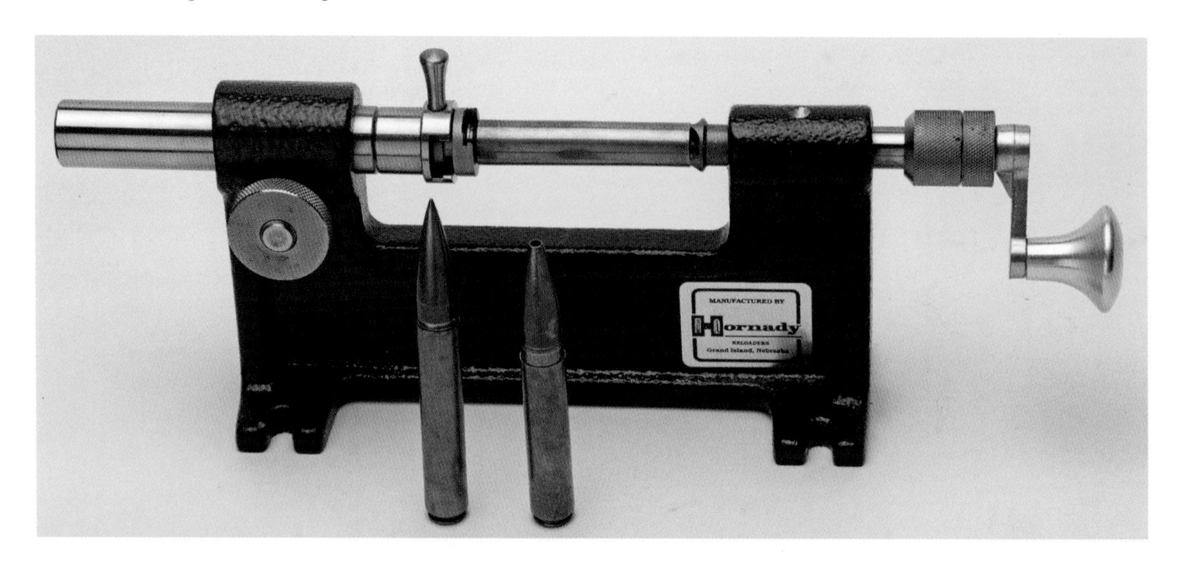

The Hornady Cam-Lock case trimmer is good for big cases like my 500 Kimera wildcat.

The Hornady Cam-Lock case trimmer is another good product used for some larger cartridge cases as it produces accurate trim lengths and allows the cases to be secured by a cam lock system with a piloted head cutter. Both benefit from interchangeable elements, so most cartridges can be trimmed to length squarely to the cutter. The cast trimmer includes the seven most popular pilots: .22, 6mm, .270, 7mm, .30, .38 and .45 calibres. Additional pilots are available for .20 calibre through to .50 calibre. An optional .17 calibre pilot/cutter is also available. Add additional tools include a chamfering tool and a de-burring tool, in place of the cutter, to perform consistent and square de-burring tasks. A power adapter is available, although I prefer hand operation.

Choosing the Correct Dies

This is another subject that has already been covered in *Sporting Rifles*, but here are a few pointers to what I use and why. The choice is between the standard $^7/_8$in 14 pitch dies used in a reloading press and the hand dies that require an arbor press to load. Both types are capable of producing superb cartridges with extreme accuracy. Threaded dies are widely used. A useful purchase is a set of full-length dies as it is usual to resize a case to its original dimensions after shooting, since this allows smooth functioning in a rifle. It should be remembered, however, that bottleneck cases need lubrication to size: if too little is used it gets stuck, but too much results in hydraulic dents to the case. Also check that the sizer button is well lubricated. This is the part of the de-capping stem that resizes the neck as the case is removed from the die and it is important to lube this for a smooth and uniform size. Dry lube is better for this as other-

Fitting an 'O' ring between the die and press allows it to self-centralize for maximum accuracy.

An arbor press allows precision-made reloads that can be made on the bench or out in the field.

wise powder can get stuck in the lube while still in the case neck. Another tip is to set the full-length die with an 'O' ring between the locking ring and the press body. Todd Kindler, the inventor of the 20 Tactical cartridge, told me this trick as it allows the die to centre in the press, resulting in more uniform sizing of the case.

For most custom projects, especially after fitting a custom barrel and possibly one with a tight neck chamber for the calibre, I would certainly go for a neck bushing die set as these only size the neck portion of the case. The remainder of the case is fire formed to fit the chamber and

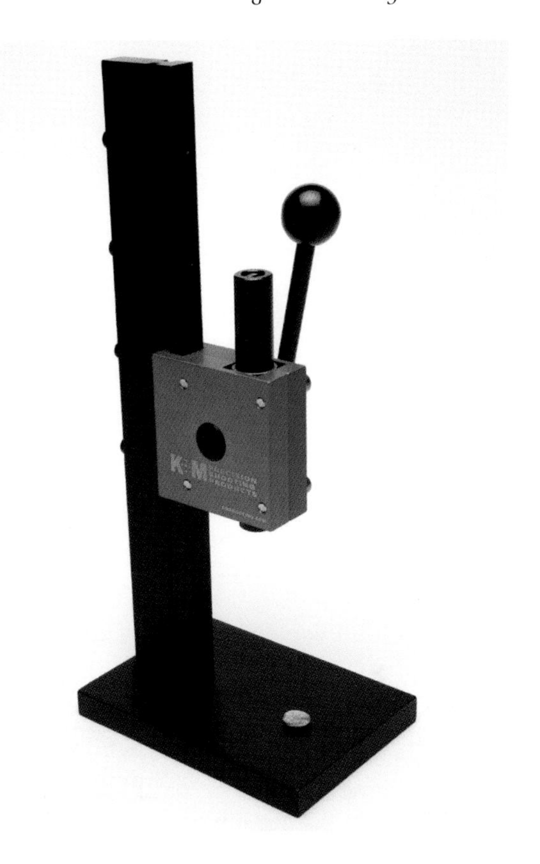

The K&M arbor press is used for precision reloads.

only a reduction in neck dimensions is needed to restore it back to normal and achieve a proper neck tension to grip the bullet. The bush can therefore be matched to the exact size you require. Most standard dies enlarge the case too much with the expander ball and so some form of adjustment is beneficial. Hand dies that use an arbor press are easy to use in the field or on the back of a pickup, although I also have a reloading cradle for standard reloading presses. I like the feel of an arbor press as you are only sizing the case neck to your custom rifle's chamber, using replacement bushes to match the neck's dimensions precisely. This makes the whole process of reloading very tactile: if a case feels sticky or has a loose primer pocket or tight bullet seating, you get direct feedback through the arbor press operating handle.

I have a Sinclair arbor press and use Wilson and Neil Jones hand dies and close-tolerance dies to produce reloads that are as accu-

Neck bushes of differing sizes are perfect for matching neck diameters on reloads and sizing of wildcat creations.

rate as you can make. For all of the wildcats I use blank dies, usually by Wilson, which are reamed out with the chamber reamer and then a replacement bush section is machined for altering the neck dimensions. The seater die has a calibre-specific seater stem added to it. A micrometer is used to measure precise seating depths that are also reproducible. You can also resize only part of the neck section, leaving the bottom portion of the neck un-sized: the sized portion grips the bullet with the correct neck tension and the rear portion allows the case to centre in the chamber, again achieving better concentricity to the bore. Neck tension too is very important, not only for a correct grip on the bullet so it does not slip in the case, but also in giving some degree of initial pressure build-up behind the base of the bullet for consistent ignition. Ordinarily you would exert a neck tension of a couple of thou (one thousandths of an inch). On a .243 calibre with tight 0.263 neck chamber, for example, this means 20 thou touching the chamber's sides, so neck turn to 0.261 on a loaded case for a clearance of one thou on each side. When necking down a fired case you need a 0.259 neck bush to produce a two thou neck tension for a 0.261 loaded round.

Sometimes, however, I like to increase this to three or four thou if the rifle is to be handled roughly on a trip aboard and still ensure the integrity of the cartridge. With subsonic rounds I have found, especially with .308 Win and the larger subsonic .338 BR, an increased neck tension is beneficial. It allows the small charge of powder, often with a magnum primer, to ignite the powder properly irrespective of the position of how it lays in the case.

Sooner or later the issue of doughnuts rears its ugly head. They are the bane of any custom project and especially when manipulating custom cases, for example with wildcat calibres. A doughnut occurs at the point where the shoulder joins the neck section of a case if there is more metal on the inside of the case than there is on the outside. This creates a small ring or doughnut of extra brass at the base of the neck on the inside. When a bullet is seated this can cause issues as there will be a corresponding hump or high spot on the outside of the case, which will affect accuracy and may cause pressure issues.

Neck turning the exterior actually makes the problem worse as it thins the outside diameter. It is best to cut out the doughnut with an inside neck reamer tool. Some cases are worse than others, but a little effort will repay the attention.

Removing the doughnut of brass at the base of the shoulder in the case with a calibre-specific neck reamer is crucial for perfect ignition.

Bullet Collimator

I have a Sinclair bullet collimator that sorts bullets quickly and accurately by base to ogive length. The weight of its black granite base allows it to be used free-standing on a loading bench. A dial indicator is attached to a secure rod and a lever control allows easy insertion and removal of bullets into the calibre-specific comparators. These can be ordered as single-calibre units (.172, .204, .224, .270, .308 and .338 calibres) or hexagonal comparators with holes bored and throated for .224, 6mm, .257, 6.5mm (.264), 7mm and 308 calibres arranged around their facets.

Insert a single round from a box of bullets to be measured in turn and any differences in length are easily indicated by the value on the dial indicator. It is very simple and accurate means of selecting bullets that are exactly the same length to improve accuracy.

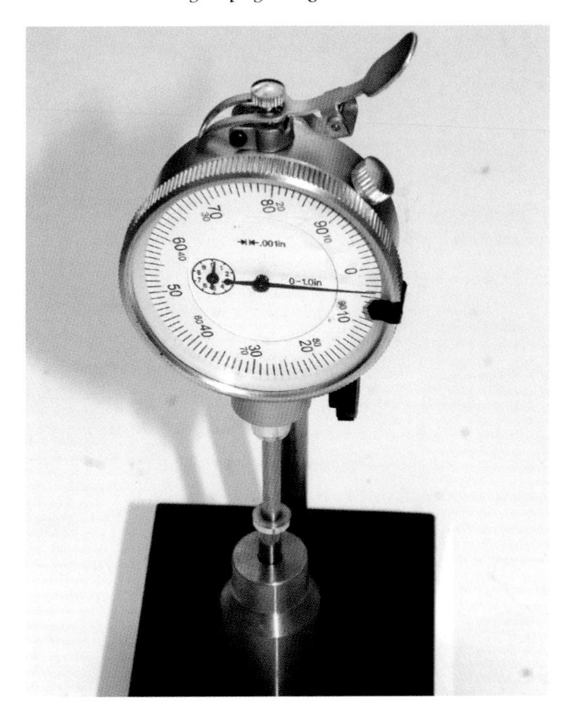

The Sinclair bullet collimator allows bullets to be sorted quickly and accurately by base to ogive length.

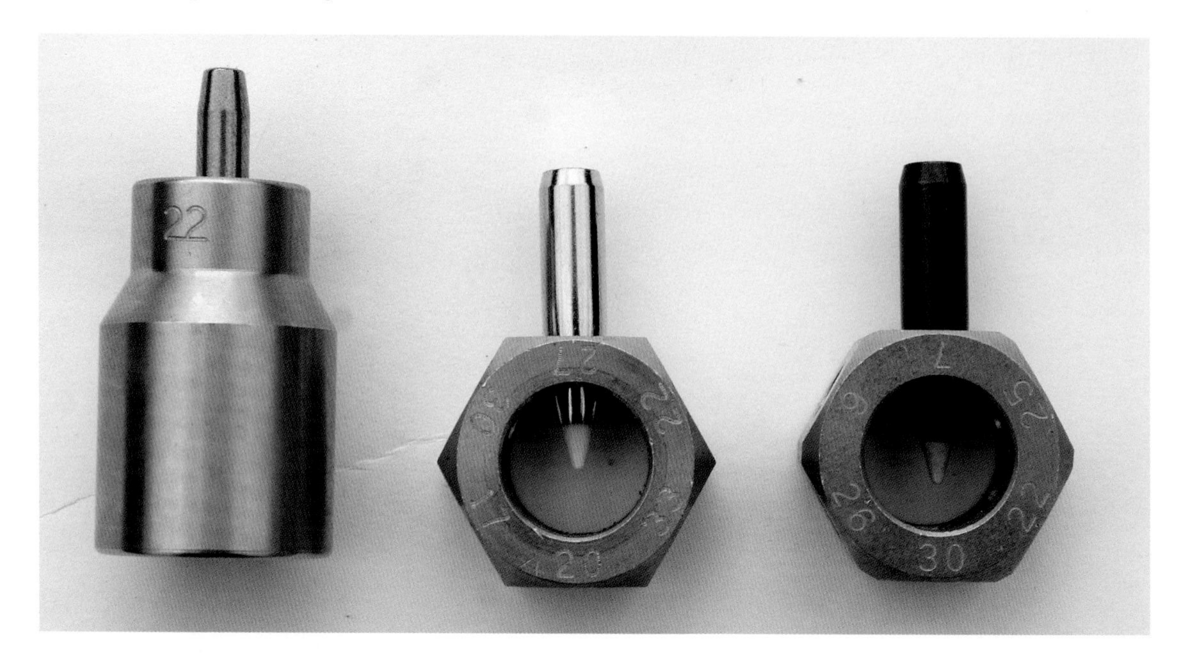

The Sinclair collimator is a simple means of ensuring that all bullets are the same length, so improving accuracy.

Hornady Lock-N-Load Headspace Tool

This headspace gauge set allows a reloader to measure changes to the headspace of cartridges with a digital caliper. Differing bushing sizes for each calibre are inserted into the comparator body, which in turn attaches to the caliper blades. You can then measure the differences between fired and re-sized cases, allowing the sizing dies to be adjusted for an accurate fit into the rifle's chamber. It is very handy if the resizing dies have been set too tight and the cases are resized too small, leading to excessive headspace and causing the cases to stretch when fired. The gauge measures cases from the case head to the datum line on the case shoulder. Five bushing sizes are available, allowing the headspace to be measured on most bottleneck cartridges from 17 Remington through to the belted magnums. These may be purchased separately, so you can choose the calibre that suits your rifles.

RCBS Precision Micrometer

This is a cartridge-specific micrometer headspace tool designed to assist in setting up full-length sizing dies in order to adjust the precise amount of shoulder bump needed. The gauge is initially calibrated using a case that has been fired from the rifle. The micrometer head can be set to display the shoulder-to-base reading needed in comparison with the initial reading. A bullet seating depth tool can also be used for the initial seater die set-up.

Calipers

Calipers are the mainstay of all reloaders and you cannot survive without them. The accurate measurements offered by these tools, which may have an old-style dial or a digital readout, are essential to guarantee accuracy and safe loads. It is foolhardy to skimp on this essential tool and hope a cheap set will suffice. The Mitutoyo digital caliper I use was expensive, but it is always completely accurate. The opening jaws and thumbwheel action run smoothly, and I find the LCD readout is easy to read down to 0.0005in (0.01mm) and is accurate to 0.001in (0.02mm).

ABOVE: *The Hornady body collimator measures the differences between fired and re-sized cases, so that sizing dies can be adjusted for a correct fit into the rifle chamber.*

RIGHT: *A RCBS micrometer cartridge headspace tool will enable you to set up full-length sizing dies correctly.*

BELOW: *A decent set of calipers is an essential part of any custom rifle reloading project.*

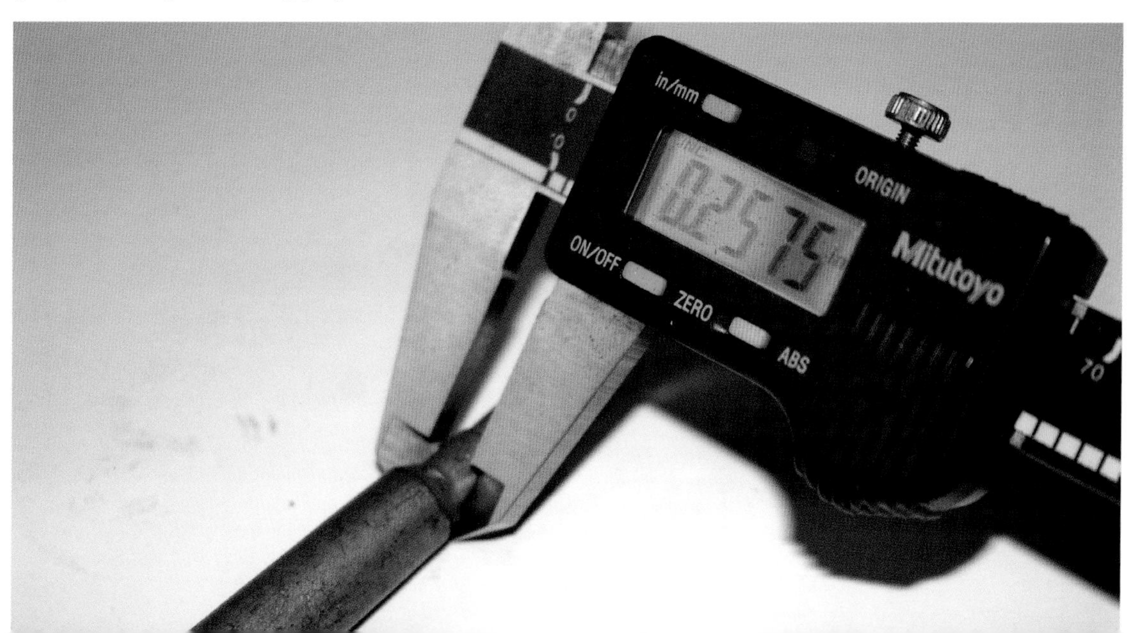

Meplat Trimmer

There has been a trend towards trimming the meplat (or nose) section of the bullet. Quite often the closure of the copper jacket on hollow point bullets can be less than uniform and it is thought that this increases wind resistance and possibly affects the bullet's aerodynamic stability. The Sinclair meplat trimmer is a hand-operated tool that accurately makes the tips of bullets uniform both for reloaded and for factory-loaded ammunition. The tool includes a calibre-specific Delrin housing, which is available separately, a tool steel cutter, an aluminium frame and a crank handle. If you were to trim off about 0.005in (five thou), this would reduce the ballistic coefficient value by only 2 per cent but it would increase consistency, so it is well worth it.

Neck Gauges

A precise set of neck thickness gauges is essential for accurate reloads. This is especially so on a custom project with a rifle shooting a wildcat or tight neck cartridge. Knowing the exact neck thickness of the case is essential to allow a safe clearance between it and the rifle's chamber: too tight and pressures can rise alarmingly, but too loose and the case has to expand more to seal the chamber end and a bullet can start its progress down the barrel off centre. The gauge also checks for irregularities in the wall thickness from one side to the other, which also can cause a bullet to cant and enter the rifling on the skew. The Mitutoyo IP65 digital read-out calipers allow a calibre-specific spigot to be attached and then a precise read-out made to check the individual cases.

A Meplat trimmer is used to make the bullet tip uniform, improving flight stability and increasing accuracy.

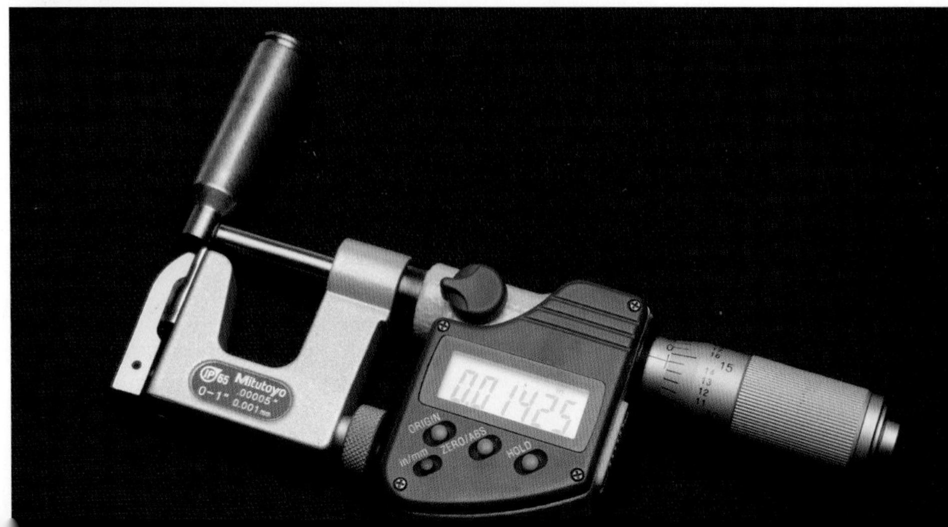

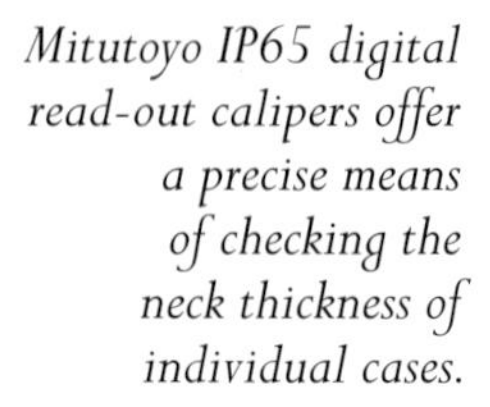

Mitutoyo IP65 digital read-out calipers offer a precise means of checking the neck thickness of individual cases.

Neck Turning

This is probably the most crucial part of any custom rifle reload, since if the rifle's chamber uses a non-standard neck dimension, for example a tight neck, you will need to neck turn. This not only removes excess brass from the case neck, allowing a proper fit to release the bullet as the cartridge is fired, but it also removes irregularities in the neck thickness resulting from manufacture or case manipulation. If the neck has more brass on one side than the other, the neck tension, that is the amount the case grips the bullet, will be uneven and the bullet will not enter the rifle correctly. This is especially important if using a wildcat round that has been necked up or down from a parent case, as then the neck might be stretched or thickened with ridges, bulges or doughnuts inside. These should be removed.

A neck turner is an essential tool for dealing with tight neck chambers or removing brass on wildcat cases.

Sinclair NT 1000 neck turner

This is the larger of Sinclair's neck turners and has a big stainless steel handle that makes it far easier and more comfortable to use. It uses a turning mandrel, made of high grade stainless steel, with a diameter 0.002in less than the bullet. This gives a perfect fit in the case, while allowing smooth rotation with the help of the lube, and avoids deforming the inside of the case neck. More expensive carbide mandrels are also available that offer superbly smooth case operation without galling. A high-speed steel tool cutter is sited at 90 degrees to the case neck to reduce the brass smoothly around the neck where the case sits on the mandrel. It is infinitely adjustable to remove the tiniest amount for a light skim from the neck or excessive bulges. The case is simply moved slowly down the mandrel's shaft as the cutter removes the brass. Where to stop is always a contentious issue, but I prefer to cut down to the junction of the neck and shoulder for a uniform total neck length.

By using an individual K&M neck turner for each calibre or case project, I can be sure that it is set correctly when I need to make more cases.

K and M neck turners

I have often used these neck turners for my wildcat loads. Although they are adjustable, once I have set it to the correct neck turning

Neck turning can be tedious, but the results obtained on this .22-284 wildcat are worth it.

depth I tend to leave it and buy another one for a different cartridge. In that way I know it is set exactly where I left it. There is a separate neck-turning pilot that is calibre-specific and is easily changed to fit inside the new case neck after it has been expanded to that calibre. This way any unwanted excess brass gathered around the neck from swaging down the calibre can be removed. This is achieved by a cutter at 90 degrees to the piloted shaft and is adjustable to cut as fine as 0.0001in at a time.

The original K&M cutters were small and your hand can start to hurt after turning necks on fifty cases or more, so the upgraded and enlarged palm grip is worth the money and makes neck turning much easier.

Expand Iron Uniformer

One tool that you should definitely have, cheap and easy to use, is an expander die with a floating mandrel fitted inside the die so that it cannot pull out and allows it to centre in the case neck. The expander mandrel is sized 0.001in less than the bullet's diameter. Mandrels of different sizes are used for differing calibres and they can be modified by a gunsmith to suit a special project or wildcat round. When entered into the case neck with a good lube, the mandrel does an excellent job of ironing out any imperfections to the neck and uniforming the neck interior for consistent neck tension. A series of different sizes can be used to neck up cases to accept larger calibre bullets for wildcat rounds.

Neck expanders or uniformers are an essential way of ironing out irregularities and making sure the case's neck diameter is uniform.

Powder Weighing

Most reloads can be adequately thrown by using a set of simple beam balances as individual loads. I use this method regularly, but most loads are now thrown from powder measures or electronic scales. Many of these are excellent, but I tend not to trust electronic scales and prefer to use a balance and precision benchrest-quality powder measure. Those made by Harrell deliver precise powder measures that can be checked on an accurate scale and then set to throw precise weights at every pull of the handle. I use two models depending on the cartridge size and amount of powder needed to be thrown.

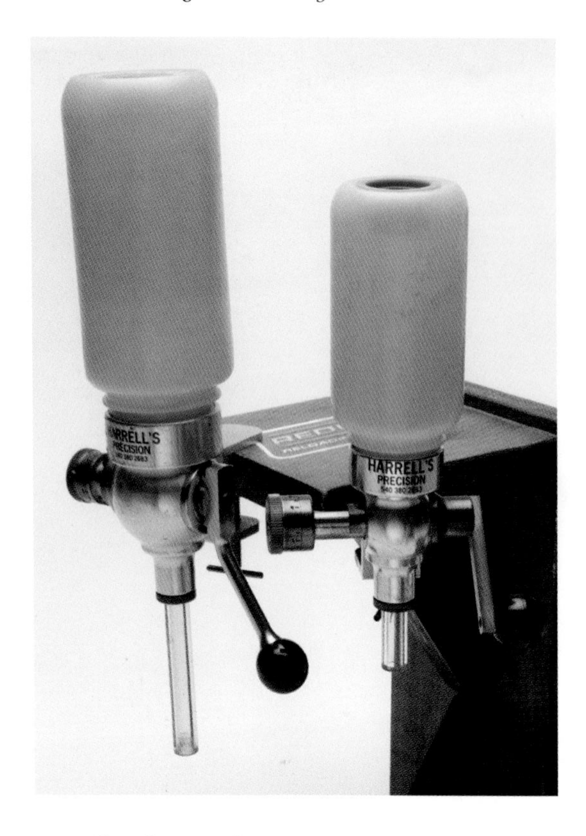

Harrell Culver powder measures ensure precise metering of powder measures when used with custom rifles.

The Harrell Premium measure can throw changes from 6 to 120 grains, so will work for nearly any cartridge you need to reload. The 16oz powder bottle holds more than enough powder and also accepts 1lb containers from manufacturers including Hodgdon and Accurate. Each Harrell measure is built with a quality CNC machined aluminium body with an extremely accurate Culver style brass metering system. It comes with a clamp-style mounting system, an appropriately sized powder bottle with a bottom plug, and two 4in drop tubes, one having an internal diameter of 3/16in and the other ¼in. Longer or shorter tubes can also be bought. This method allows the case to be filled more accurately as the flow of the powder grains is more uniform, packing the case more evenly and so allowing fewer air spaces. As a result it will burn more consistently. This is rather like the result of the swirling technique when using a powder scale pan to deliver powder into a case, allowing the powder to swirl in one direction around the powder funnel. Both techniques are valid and beneficial.

The Premium measure has needle bearings built into each side of the metering insert. It is very smooth to operate and you can really feel the powder being measured and thrown, so any hang-ups from long-grained extruded powders are obvious. Short cut or ball powders are more easily managed in any powder scale. These measures are expensive but they throw with the same accuracy every time and I really like that you can dial in a number on the scale that represents a particular powder charge and know that it will be accurate each time. Many of my reloads are referred to in my reloading data logs as 50H or 25H plus two clicks, referring to a particular powder weight. I find it easier to use this method. For my 6mm PPC custom, for example, I know that every reload will be 50H of Vit N133 powder with Hornady 65gr V-Max bullets, without having to weigh out individual powder charges.

The second Harrell I use is the small Schuetzen/pistol measure, which is superb for delivering very small, accurate loads for small-calibre wildcats like the .14 Walker Hornet or .17 Squirrel. It is also indispensable as an accurate measure for subsonic calibres from 300 Whisper, 300 AC Blackout to .338 BR/Whisper and all my .308 Win subsonic loads. Each click on the Culver measuring system accurately changes the powder charge by 0.03 grains. This model has a powder weight range from 2 to 25 grains, and uses the smaller 250ml (8oz) powder bottles.

Chamber Length Gauge

This gauge is used to determine the exact length of the rifle's chamber when measuring from the bolt face to the end of the chamber neck. This instantly tells you the actual length of the rifle's chamber length, which is not always as stated on the reamer print or manufacturer's specification. The gauge is made from 12L14 soft steel and is calibre specific. It is placed into a shortened, once-fired case, inserted into the rifle's chamber and the bolt is closed. The gauge then slides down into the case as it touches the true

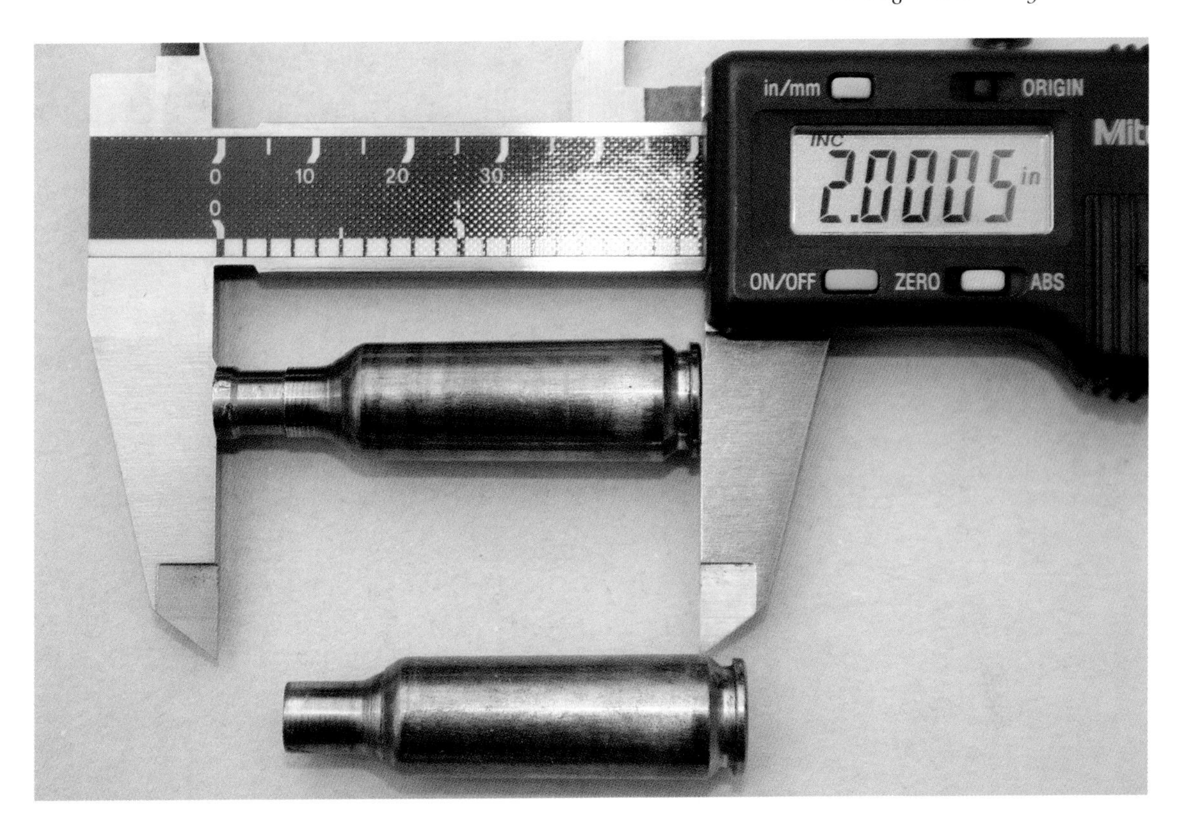

Chamber length gauges are cheap and handy little tools for measuring the chamber length of your rifle so that the cases can be trimmed precisely to size.

chamber's end. Take the gauge out and measure it to find out the real chamber length as an accurate guide for trimming the overall length. A case that is too short will never shoot at its maximum accuracy potential, while one that is too long will cause pressure issues.

ANNEALING

A very important step in any custom case preparation, especially if you are feeding a wildcat chambered rifle, is to anneal the brass. If you have to neck down a .30 calibre case to a .22 calibre for a super-fast varmint round, then the process of necking down or expanding the brass will work harden it and make it brittle. If you anneal the case it re-uniforms the metal molecules back to normal and improves the case life, and it will certainly reduce the risk of split

Working brass when resizing can make the case brittle. Annealing realigns the brass on a molecular level for longer life and improved uniformity.

necks or separations. The process is very tricky and not an exact science, and it takes some practice to get it right. I anneal my cases manually using a Woodchuck Den annealing tool, which employs a uniform copper tube profiled into a ring so you can encapsulate the case to be annealed from all angles and get an even heat to the neck and shoulder area.

Cases sit in a bath of water to stop the head section of the case overheating as this will cause the case to fail. I usually heat to a dull red colour and then topple the case over to cool in a bath of water and repeat the process. It is one of those jobs that seem a chore but can make a significant difference to case life and performance.

PRECISION RELOADING FOR PERFECTION

Load Development

Following all your hard work sorting, manipulating, turning and measuring cases, you now need to develop an accurate load in a methodical way to maintain consistency, otherwise all the time is wasted. I initially use the QuickLOAD ballistic program to determine a safe load for the dimensions and barrel length of my custom rifles. Then, drawing on a little finesse, I develop that load further to fine tune it for my rifle's own idiosyncrasies. You are trying to achieve the smallest group size possible to maximize accuracy downrange. In load development it is good practice to change one factor at a time. You may start, for example, by altering the primer you use but keep the powder type, weight, bullet and seating depth the same, because that way you can see if the change actually makes a difference. The next step would be to change the powder type, and so on, until you are happy with the results. If I start with a reload from QuickLOAD, I then alter the load by changing the powder charge at incremental points below the maximum pressure level indicated, as accuracy from a barrel is never usually at full throttle.

As discussed in Chapter 3, barrels have a resonance to them when fired. I start by changing the powder weight by ½ grain increments (sometimes ¼ grain changes for smaller calibres). What you will notice from the group sizes is not only a change in overall size but also the orientation of the group, either vertical or

Getting a wildcat or reload to shoot can be a tedious task, but it is all part of owning a custom rifle.

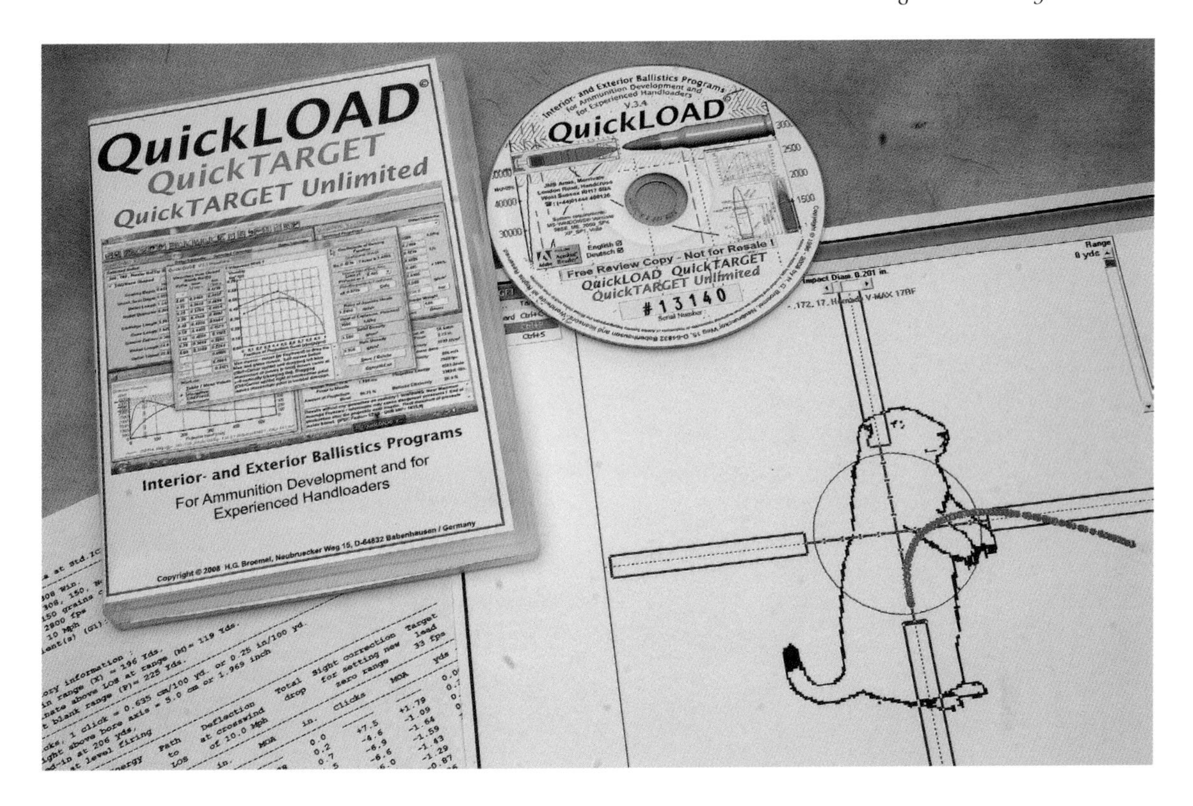

When you want to predict a cartridge's performance without shooting it, QuickLOAD and QuickTARGET save time and components.

Load development is just that. Even a small change in the seating depth of the bullet can have an effect on its accuracy.

horizontal. This is because every load puts different vibrations through the barrel and so the harmonics, as the bullet travels down the barrel, will alter with movement at the muzzle end, so influencing the accuracy. With load development you are trying to achieve a load where the bullet exits the barrel when it is in exactly the same orientation, shot for shot. As the barrel moves in its up-and-down wave arc, you want it to be in a state of least vibration or movement at the point when it changes from one position to the other. This is called barrel timing, as illustrated by the load development information shown here relating to a .20 BR load. Once the

A Sinclair priming tool is used to set the primer perfectly in place into the case. The vital importance of primers is often overlooked when seeking means to improve accuracy.

most accurate load has been achieved the data can be entered into QuickLOAD and it will predict all the alternative powder types that will replicate the barrel timing.

Every cartridge has a sweet spot of harmonization with a particular load of bullet, powder and primer. With the most accurate load determined with a particular powder, the primer can now be changed to see how this affects the group size and muzzle velocity. A small, but sometimes significant, part of reloading for a custom rifle that people forget is that primers burn at differing temperatures and, more importantly, over differing durations of time. This significantly affects the burn rate and inside turbulence of the powder combustion in the case, so it is worth the effort to change the primer with a view to achieving consistency and total efficiency each time. Match grade primers are worth buying as they burn strongly and consistently. For most subsonic loads in larger cases I would recommend magnum primers to achieve the correct burn of the powder. It is simple to buy a selection of primers and modify this part of the reloading procedure, seeing how it changes the velocity and horizontal spread to groups as it adjusts the barrel timings. Precision primer seating tools, such as the hand-held Sinclair priming tool, can be used to ensure the primer is perfectly seated in the newly reamed primer pocket and achieve consistent primer seating depth without distortion.

Another procedure that can make an enormous difference to the accuracy of the load is to alter the seating depth of the bullet. For more details on this, please see *Sporting Rifles*.

Load Density

This is a very important part of producing an accurate load as the way a powder charge sits inside the case will affect the way it ignites and burns. The gaps or air spaces between the powder grains can vary enormously depending on whether a ball powder or extruded type powder is being used: a 100 per cent load density stops voids forming in the case and affecting an otherwise efficient burn.

You will find that changing the way powder is dispensed into the case can also alter the load density for the same charge weight. The use of differing lengths of drop tubes that is possible with Harrell Culver measures allows the powder to travel in a manner that to some extent is uniform, enabling it to enter the case in a better packing state. I also place my thumb over the top of the case and tap it on the loading bench to further settle the powder in the case. After a while this will become second nature. If you use a beam scale and load more individual charges by hand, as in my .14 Walker Hornet, then the swirling technique helps to pack a case correctly without voids. If you allow the powder from the beam scale's pan to pour slowly on the edge of the powder funnel in the case neck, and not straight down the centre hole, you will find the powder arranges itself in a more orderly manner.

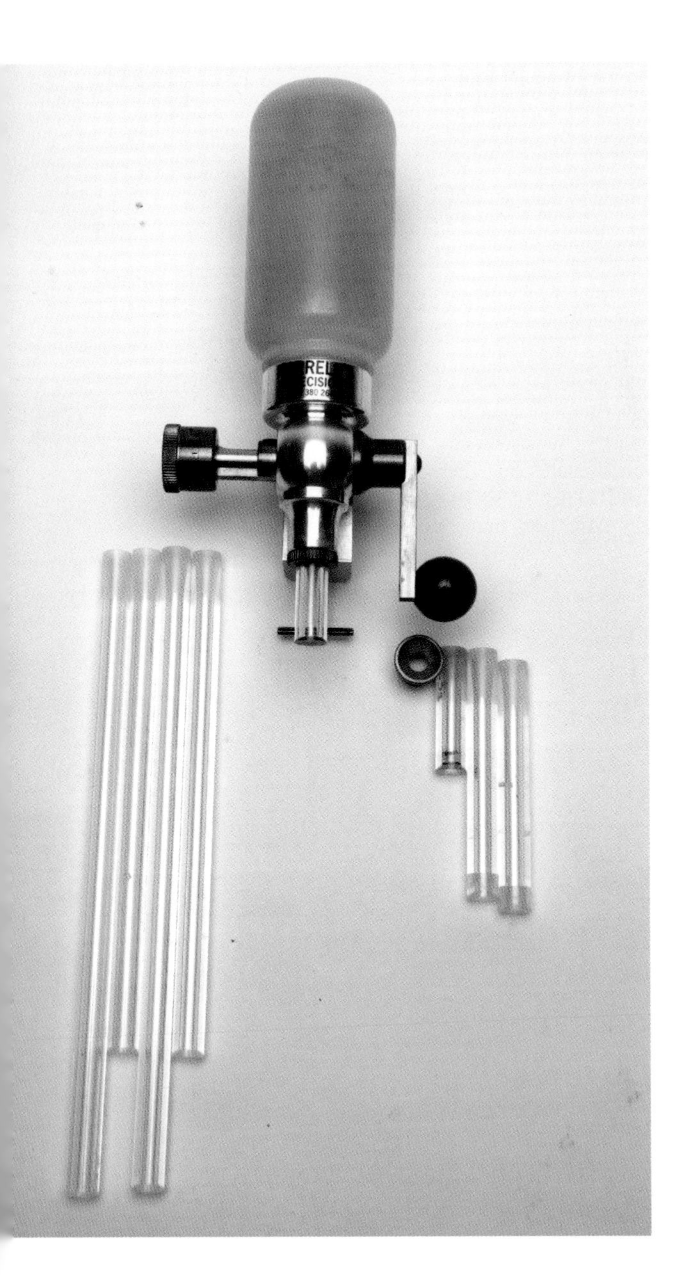

One way of ensuring a perfectly filled case is to use drop tubes that are a different length to my Harrell powder measures, pouring with a swirling technique.

WILDCATS

Wildcats – love them or hate them ... Do we need them or are they a waste of time? It all depends to some extent on your views and needs as a hunter. A standard factory round is more than sufficient for the vast majority, and that is fine. My favourite calibres are .308 Win and .22LR, but without development and experimentation these two great rounds would never be in existence. They all once had their origin in a wildcat: the difference is that most factory loads are just the last development of a series of wildcat rounds that then become commercialized. As with so many things in life, people like to push the boundaries, experiment and improve almost anything to achieve a product that is better and truly different. Whether this is for your own personal satisfaction or derived from the need to succeed financially, the end result is still the same, a sense of being part of the process and owning something unique.

What is a Wildcat?

A wildcatted round is one that has been modified to improve its performance in terms of accuracy, velocity and case life. You usually start by reforming an existing cartridge case. This may involve removing any body taper and increasing the shoulder angle in order to create a higher powder case capacity and therefore velocity. It may just require necking up or down so that larger or smaller projectiles can be shot from the parent case. It is true that modifications like these can seem daunting, but in reality most shooters are already shooting some sort of wildcat. Many of the great cartridges available on the market, such as the .22-250, .243, .270, .25-06 and .280 Rem, are standardized and commercialized wildcat rounds based on the parent .30-06 Springfield cartridge. The list is endless. Although all wildcats utilize different combinations of cases and consumables, there are certain criteria you must follow to achieve your dream rifle.

You must decide whether to follow an existing wildcat round route, for example in the steps of the 6mm AK, .22 Cheetah or the 500 Whisper, or strike out in a different direction. Choosing an existing wildcat takes away much of the guesswork as reamer prints, load data and reloading technique have been explored before. If this is your first wildcat venture you should start here. The more adventurous can start bending brass to create their own wildcat. If you take an existing .308 Win case, for exam-

Wildcat cartridges are fun to make, design and shoot. The results are not always an improvement, but the enjoyment comes from experimenting with something different and learning about what affects a rifle's performance and ballistics.

Choosing the best components and reloading tools for feeding and making a wildcat is a fascinating hobby that can result in some interesting reloading kit.

ple, shorten it, lengthen the neck and improve the shoulder angle, what you have in essence is a wildcat, even though the dimensions may differ from the standard .308 Win by only a few thousandths of an inch, and you can legitimately call it a new name. The .204 Ruger, for example, was originally the .20 Terminator. For the more experienced, a complete calibre change is what keeps wildcatters striving for more. It can be intriguing to follow the path from the 6.5mm Panther to the .260 Rem, .260 Rem AK Imp, 6.5 Creedmoor and the similar 6.5 × 47L, all of which share much but have their individual differences.

How to Get Started

It all starts with a particular bullet that takes my attention and curiosity whether it can be pushed at a particular velocity. Usually this means a high ballistic coefficient bullet for better downrange performance. Then I see what is out there in terms of ballistics from existing cartridges and if there is any room for manoeuvre. It is getting harder, though, as most cases have been wildcatted in some form or other.

I would love the luxury of making my own dimension cases but that is just not economically viable, which is why a parent case is used. Take the .20 Satan wildcat, for example. I love small calibres and jumped on the 50 grain and later 55 grain bullets when they were introduced, resulting in a better ballistic coefficient than the standard .224 version for the same weight bullet. You need to look for a case that offers opportunities, but that is easier said than done. If a case is too big you can say goodbye to the rifling due to excessive wear, but if it is too small, there is no real advantage. I use the ballistics programs QuickLOAD and QuickDESIGN to design a new wildcat from the existing standard cartridge database. When it came to the .20 Satan I wanted to use the 6 × 47mm Swiss Match and later the 6.5 × 47L cases as donor cartridges. With the actual cartridge cases in front of me, I used the program to redesign them into a new form with which I was happy. By changing or completely scratch building a case in QuickDESIGN, your new creation can be modified at will in the virtual world. An alternative method is to use existing reloading or form dies and making the neck bushing sizes either larger or smaller, depending on the final bullet diameter, to swage the neck. Then use alternative dies to reform the body or shoulder angle to get a partial wildcat. This way you can visualize a prototype in real time. With an initial new case design, I weigh the case and measure the water capacity to check internal powder capacity. I then run these dimensions through the QuickDESIGN program to see if I am close to my virtual case. Differing manufacturers' cases vary a lot. Now you need to decide whether you want a tight neck or standard neck dimension: a thinner neck will need neck turning to gain more precision and the correct neck tension.

.22-284 wildcat, also known as .223 Valkyrie, is a superb ultra-fast barrel burner that can achieve more than 5000 fps with a 30 grain Berger bullet.

With the dimensions sorted, I shoot it onscreen in QuickLOAD with a variety of bullets and powder combinations. This calculates very realistically the true ballistics your new creation

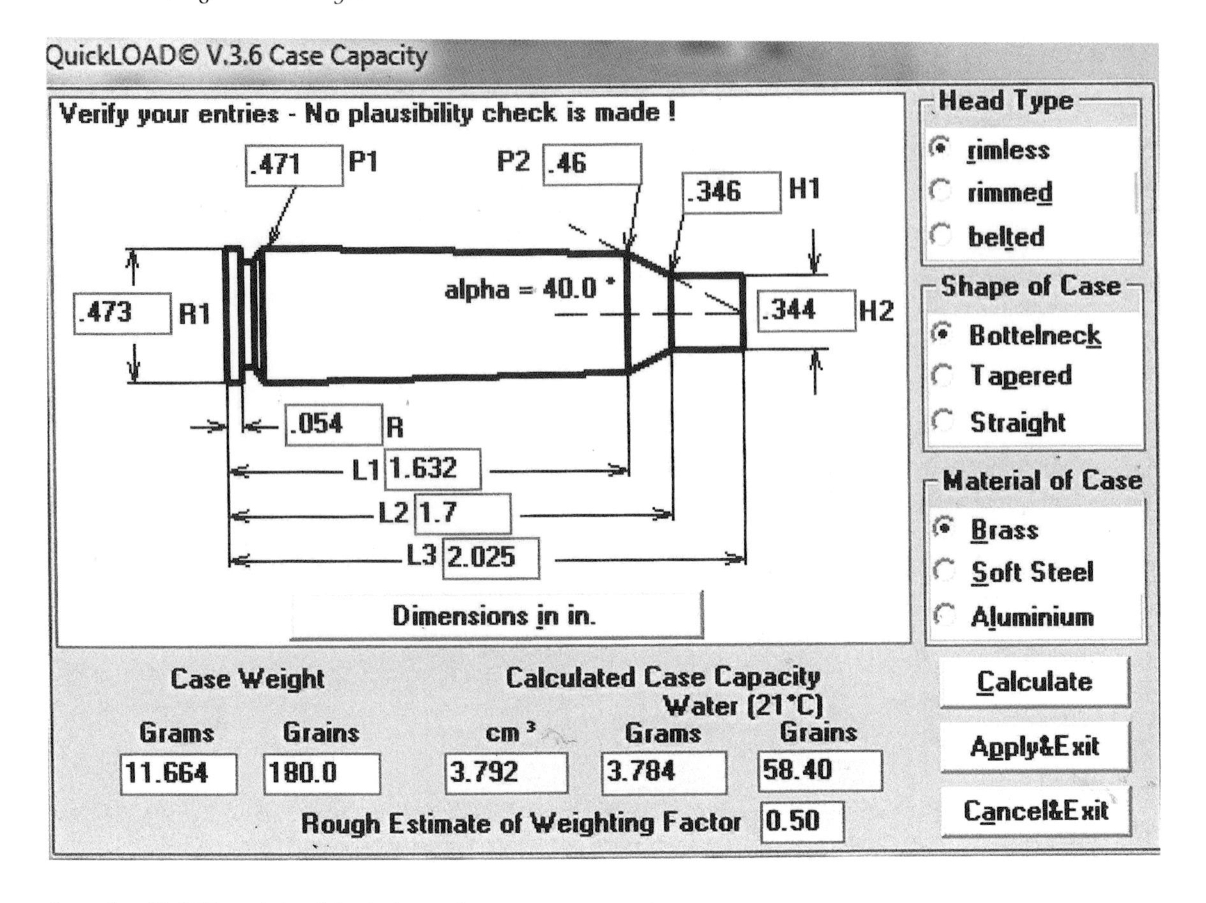

I use QuickLOAD and QuickDESIGN to design my wildcats, fine tuning a reamer print at home before sending it to a reamer maker.

may achieve. It is here you cross your fingers and hope the pressure scale stays safe and the velocity figures correspond to your initial predictions. If it doesn't work, go back and re-compute to change a few dimensions. Smaller changes are better in ballistics, as a small change can make a big difference.

I now print off a case design sheet as a reference to give to the reamer maker. Include as much information as possible: not only does the chamber reamer need to suit your new design, but you have to decide on seating depth, and therefore throat length and neck diameter. If I use a printout I tell the reamer maker the bullet I am shooting and suggest an overall length of the loaded round, based on the ogive of the bullet touching lands. If the case is made from a parent cartridge, I send three copies to the reamer maker as a dummy round. You then have to wait for the reamer to be made. I use a reamer with a rotating pilot, so each pilot can be changed to get a true fit in the barrel. I also order a go gauge to check headspace.

Now you need to choose a barrel with the correct twist rate to stabilize the selected bullet weights and the number of rifling grooves. The next consideration is overall barrel length, the material it is to be made from, such as stainless steel or chrome moly, the contour and whether or not it should be fluted. Formulas are available to ascertain the best twist rates, but when shooting a wildcat you have to consider barrel erosion or any extra pressure resulting from too tight a rifling twist. Other considerations should include whether three lands is better than six, should the leade be 1.5 or 3.0 degrees, and whether you want ratchet, polygonal or normal flat rifling. You will also need enough length to achieve the extra velocity you hope your new creation will deliver. At this stage it is an unknown entity, so you may just have to wait and see. Once your barrel has been newly reamed, chambered and fitted, the fun can begin.

If the wildcat is a simple neck down or up you can fire-form your load in the chamber to its new dimensions, usually with the bullet in the lands to stop case creep but using a mild load. This means that although the case is formed using mechanical dies, it still needs to be shot in the chamber of the rifle to expand it to the correct proportions. For this you will need some resizing dies and a seater die to reload your cases. If it is a known wildcat, it is possible that a set of dies may be available, for example from the Redding Reloading Equipment custom shop or the die makers CH4D in Ohio.

I sometimes use existing dies, cutting them down or modifying them so I can resize the neck or body, or even seat bullets in the new wildcats: with the 300 Broadsword round, for example, I used 30-47L, 300 WSM and 308 × 1.5 dies to create a finished round. When working on a one-off where the neck and length dimensions and the shoulder all differ, however, you will need a set of dies tailor-made to your case. I tend to use blank dies from Wilson or Newlon and then have the gunsmith use a chamber reamer to cut a two-die set. The first will be a neck die, where either Wilson or Redding neck bushes can be used to achieve correct neck tension. Using the correct seater stem, the other die can be used as a seater. If you want a full-length die you will need a smaller reamer dimension than the one that cut the chamber, or sometimes you could use a body die that may fit the new case.

On first inspection all this hassle may seem laborious and an unwarranted expense, but without the experiment involved there would be no advances. Some wildcats deliver only a meagre 5 per cent velocity gain but, as with the .308 Win AK, the new look can be fantastic. Then there is the .250 AK, which is well worth the performance increase and encapsulates the wildcat ethos in a nutshell. Like most things in life, the best results come from maximum input and effort. How boring it would be if we all drove around in Ford Escorts. People have always customized and improved what exists and that is what keeps me interested.

The pay-off from choosing the correct dies, bushes and neck tensions will be good groups downrange, such as this .22 Satan wildcat group.

MY CONTRIBUTION

What will motivate many to achieving their own wildcats is the pure enjoyment you get from forming cases, load development and all the rigours of case preparation to achieve something really different. True, many will find this too time consuming, expensive and of no benefit to their style of shooting. Why improve a .223 case for more velocity gain if you only shoot foxes at 100 yards? I, however, was influenced by my first gun magazines in the early

My small contribution to wildcats: (left to right) .20 Satan, .22 Satan, 6.5 Rapier, .30-47L and 300 Broadsword.

1970s, which were full of wonder calibres and wacky cartridge names such as the .240 Cobra, .17 Pee Wee or .22 Eargesplitten Loudenboomer! What follows is an account of how the specialized loading tools described above can be employed for peak performance.

.20 and .22 Satan

.20 Satan

Fast-moving small calibres are capable of superior downrange performance, primarily due to the enhanced ballistic coefficients of the bullets as compared to a .224 or .243 calibre projectile of similar weight. That is some statement but small calibres have enjoyed a renaissance of late. I love my .14, .17, .20 and .22 calibre wildcats, but my best to date has to be the .20 Satan. Whereas the 20 Tactical is great for bullets up to 40 grains in .20 cal, the 20BR and 20-250 certainly are more efficient when pushing the heavier 50 and 55 grain projectiles. A .20 BR round will push a 50 grain bullet at about 3850 fps, which equals the velocity of a .22-250 or .220 Swift but uses a better and higher ballistic coefficient bullet than the .224 bullet (more than 0.3 as compared to 0.25). Downrange performance will therefore be correspondingly better. The 55 grain Berger bullets are as wind slippery as they get and have a great ballistic coefficient of 0.381, which is better than a .243 Hornady 87 grain Match King bullet.

The only compromise with small calibres is that as the bullets get heavier they become longer and therefore need a faster twist rate to stabilize. This is not a problem, but if you want to push a 50–55 grain bullet at high velocity down a fast twist barrel, pressures can rise alarmingly. A 50 grain varmint boat-tail bullet needs a 1 in 9in twist to stabilize it rather than the 1 in 12in common to the lighter .20 calibre projectiles. The 55 grain bullets prefer a 1 in 8in or 1 in 8.5in twist. My rifle and cartridge specification for the .20 Satan was a Pac-Nor super-match grade barrel custom profiled at 29in of actual rifling, plus the chamber and the twist rate needed to stabilize a 50 or 55 grain Berger .204 projectile. I commissioned a new rifling button from Pac-Nor with a tight 1 in 8.5in twist rate just to make sure the 50 and 55 grainers would be stabilized. The barrel was then fitted to my RPA Quadlite switch barrel action, which Steve Bowers made for me.

The .20 Satan was my first wildcat. It shoots the new .20 calibre at full throttle and with superb accuracy.

Necking down the 6mm Swiss Match brass to .20 calibre is no problem and the results will benefit any vermin shooter.

I really wanted a three-quarter length .243 case, but did not want the chore of resizing and forming brass, so I hit on the idea of using the 6 × 47 Swiss Match case. This is a match quality .308 bolt face cartridge with a cartridge length of 47mm and therefore represents a three-quarter length .243/.308 case capacity. This was actually the mark 1 version: as soon as the Lapua 6.5 × 47mm case came out I necked this down to .20 calibre and this became the .20 Satan mark 2 version. Both are on the QuickLOAD/QuickTARGET ballistics program. In reality both cases perform very similarly, despite the 6.5 × 47L using a small rifle primer, rather than the large rifle primer of the Swiss Match case. The design and water capacity of 44 grains looked feasible and I settled on a neck diameter on the reamer of 0.235in, which would give 1.5 thou clearance either side of the neck and chamber wall with a case length of 1.8215in and a shoulder angle of 30 degrees. Just in case the load density was not optimal, I also designed an improved case with a 40- and 45-degree shoulder and a small capacity case called a Pup. Forming the brass was really easy. Virgin 6 × 47 Swiss Match brass was sorted by weight for consistency and batched accordingly. These were then run through a .22 BR full-length die set with sizing button removed to initiate the first neck sizing procedure, being sure to lube with Imperial Sizing wax as you go. I next ran the cases through a .20 BR FL die with button removed to achieve the primary .20 neck calibre diameter. I then trimmed the cases on a Wilson neck trimmer to 1.8215in and ran each case through a .204 expander mandrel to even out the neck concentricity and give enough neck tension for the bullets for fire forming. I de-burred the inside and outside of the neck and now had a .20 Satan cartridge ready for fire forming.

Medium burn-rate powders, such as Alliant Reloder 15, RL17, Varget or H 4350, are good choices. Therefore a load of 37.0 grains of RL 15 with a 50 grain Berger bullet ignited with a Federal 210 primer would shoot 0.25in three-shot groups with less than 15 fps variation and a staggering 4018 fps average velocity. That is pretty good.

The smaller calibre 55 grain .20 cal Berger bullet has a ballistic coefficient advantage over the .22 cal and .243 cal bullets of the same weight.

The .22 Satan was the next step to increase range with both the 6mm Swiss Match and 6.5 × 47L cases.

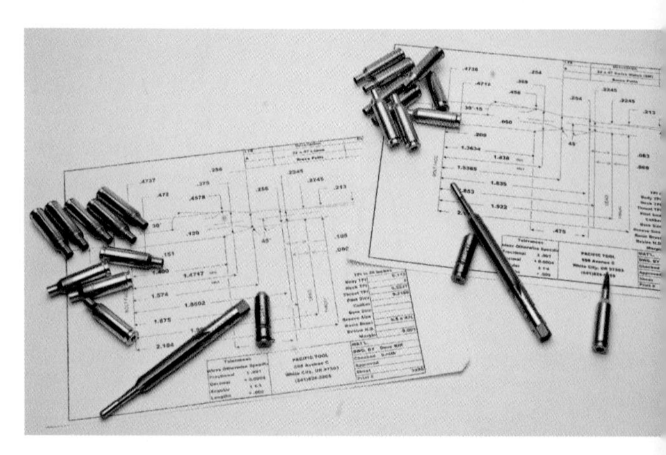

Finalizing reamers and prints for both versions of the .22 Satan wildcat was time consuming, but QuickLOAD and QuickDESIGN made the process easier.

The 55 grain Berger bullets, or torpedoes as I call them, are 0.915in long and extremely sharp and slender with a boat tail and, best of all, a ballistic coefficient of 0.381. Reloder RL17 is just right for the 55 grain bullet. I started at 34 grains, which gave a sedate 3507 fps, so a grain more at 35 grains upped the velocity to 3598 fps. Raising this to 36 grains increased the velocity to 3691 fps, and 37 grains produced a healthy 3811 fps. Varget topped out at 38 grains and 3901 fps, with H4350 achieving 3889 fps with a 40 grain payload of this powder. This means it has a good downrange ballistic coefficient advantage compared to the .224 or .243 bullets. Similarly compare the 50 grain .204 Berger bullet to that of a .224 Berger match bullet. The former has a ballistic coefficient of 0.296, while the traditional .224 bullet can manage only 0.241 (even the venerable Nosler 40 grain Ballistic tip can manage 0.242). Even better is the astonishing Berger 55 grain .204 bullet. This has a ballistic coefficient of 0.381, whereas the same weight .224 bullet is only 0.267. Even a .243 bullet of 55 grains has a low 0.240 ballistic coefficient and if you launch these at the same velocity the smaller .204 bullet will outperform the rest by a long shot, literally.

.22 Satan

The .22 Satan, the big brother to the .20 Satan round, is devilishly good. I wanted to achieve better downrange ballistics, but this time with a heavier bullet and not by sheer speed. I hoped this would mean less barrel wear and so I upgraded the .20 Satan wildcat by expanding the case to accept .224 bullets.

I originally used the 6 × 47 Swiss Match brass for the .20 Satan and now the .22 Satan, but Lapua later produced their 6.5 × 47L, case which does not have the same dimensions as the Swiss Match. The Lapua has a shorter neck but a longer body, but it does have a small rifle primer pocket whereas the Swiss Match brass has a large primer pocket. Wishing to utilize both brass types for testing, I had Steve Bowers make a forming die that would form either brass type straight down to .22 × 47 Satan dimensions. This really involved only a neck swage for the Swiss Match cases, but a shoulder bump and neck reduction for the Lapua brass. Having run through the form die, I trimmed to 1.8435in length, which was 25 thou shorter than the chamber dimensions of 1.8460in – a little short, but safe. Water capacity is then 46.5 grains for the Swiss Match case and 46.12 grains for the Lapua brass. The Lapua brass is heavier and has slightly thicker walls. After de-burring inside and out with a Wilson de-burrer, the Swiss Match brass was ready to take bullet heads: the Lapua, however, still needed a slight neck reduction to provide adequate neck tension. To fire form the brass, I used a load of 34 grains RL19 under a 69 grain

Wildcats can be expensive to make as so many unknown factors may need to be overcome, but the success of groups like this from the .22 Satan shows that it can be worth it.

Hornady match bullet, all ignited with a Federal 210 large or 205 small rifle primer, dependlng on the case used.

A Pac-Nor stainless match grade 28in barrel with a 11-degree crown and a rifling twist to stabilize 75–80 grain projectiles of 1 in 8in twist would be fine, but bullets with the 90 grain weight are rather long, so 1 in 7in or 1 in 6.5in are better. (I stuck to 1 in 8in.) A powder choice of RL 17, 19 and H4831SC would seem logical, as QuickLOAD confirmed. For varmint shooting I use H4831SC powder of 41.5 grains behind a Hornady 80 grain A-Max bullet, which can achieve 3340 fps velocity and 1902 ft/lb energy. An alternative good load would be 35.5 grains of RL19 powder and 90 grain Sierra Match King bullet, which travels downrange at 3040 fps velocity to achieve 1847 ft/lb energy. Both loads shoot 0.25in groups at 100 yards and less than a half minute of angle downrange. These loads also seem to work well where it is legal for small species of deer. A load of 43.5 grains of H4831SC powder and a Nosler 60 grain Partition bullet can achieve 3618 fps and 1744 ft/lb. Switching to RL19 powder and a payload of 42.5 grains and a Sierra 65 grain Game King bullet will yield 3631 fps and a healthy 1903 ft/lb energy, ideal for roe deer in Scotland or muntjac or Chinese water deer in England.

As mentioned earlier, I was curious to know the difference between using Swiss Match cases and reformed Lapua brass and their differing primer sizes. As it turned out, there was little detectable difference. If you want to take the easy way and avoid a few forming steps needed for the Satan brass dimensions, then Dave Kiff at Pacific Tool and Gauge in Oregon has reamers for the .22-47 Lapua version (Satan mark 2), which only need a neck reduction and all the data from the .22 Satan will be relevant. I have reamers for both, but elected to stay with the original mark 1 Satan version.

6.5mm Rapier

My goal here was to have as efficient a cartridge design as I could, while using an existing parent case of quality that would be in a deer calibre but could also double as a fox round. It was also to burn no more than 40 grains of powder and provide good velocity from a short 18in barrel. The calibre choice of 5.5mm was chosen due to my liking for the .260 cartridge and my regard for the venerable old 6.5 × 55. Bullet weight choice in these calibres spreads the range for fox and deer work and so this made sense. Lapua 6.5 × 284 parent cases were utilized as the rebated design would be perfect when reformed to a more squat format with a highly efficient internal combustion chamber.

All the primary work on the case dimensions was carried out in QuickLOAD and then Steve Bowers made the custom forming dies to reduce the brass sequentially in stages to the final dimensions. This data was then sent to Dave Kiff at Pacific Tool and Gauge to have a barrel chamber reamer made with the correct throat for a 120 grain bullet. The barrel was a Walther match grade stainless steel make with a Bowers custom profile, which made it fairly heavy, with a total length of 18in without the chamber, and a 1 in 8in rifling twist. In this way, with a muzzle diameter of 0.84in,

The 6.5mm Rapier round was designed to make a highly efficient 6.5mm wildcat from the .284 case for use in a shorter barrel for deer stalking in Britain.

Forming the 6.5mm Rapier cases was difficult as it involved lots of case reductions and neck turning. This was made easier with a set of custom dies machined by Steve Bowers.

the weight of the barrel would give stiffness as well as combat overheating and heat drift. The rationale behind the Rapier was to modify the parent 6.5 × 284 case, which is 2.160 in long with a 35-degree shoulder and burns 49 grains of powder to push a 129 or 130 grain bullet weight at 2850fps and 2327ft/lb energy from a 24in barrel. This is a fine cartridge in its own right, but it is known for shorter than average barrel life. I wanted to achieve more realistic powder consumption while increasing the barrel's longevity. The intention was to turn out a cartridge that would send a 120 grain 6.5mm bullet out of that short barrel at 2800fps and develop 2090ft/lb energy.

This would mean taking the parent Lapua case through a series of four forming dies to reduce the shoulder and neck junction gradually back to its final position, while reforming the shoulder angle and leaving a long neck for trimming. The Rapier design utilizes a case with a long neck of 0.337in, not only for bullet support but also to ensure proper powder burning within the case for as long as possible. After the third die the case is trimmed to a length of 1.88in and de-burred inside and out. The neck is then expanded with a mandrel, starting with a 6mm and then a 6.5mm size.

The next step is the neck turning, which is essential to remove excess neck brass. In this cartridge, one side is 24 thou thickness, so at least half that needs to be removed to give a good neck tension. Because the chamber dimension of the reamer is 0.2904in, I needed to turn the necks to allow a clearance for bullet release of at least 2 thou, that is with a thou either side as a minimum. I therefore neck turned to 0.287in. With a 120 grain Sierra ProHunter bullet seated in the case the outside diameter of the neck was now 0.288in, so meeting the minimum requirements. The fourth die is cut with the finishing reamer and is used as a true up die. Together with several shell base holders of different sizes that Steve Bowers also ground, this allows the die to be used as a bump die. This is very handy when the cases get sticky after a few firings. All that is left is to fire form the brass in the rifle's chamber, followed by a 6.5mm Rapier round. This, however, still leaves the dies necessary to reload the fired cases. Steve made these as well, based on the Wilson die blanks that Steve re-chambered and honed to perfection so that each reload was as consistent and accurate as possible. The relatively confined nature of the powder burn chamber inside this case and a charge weight of 50 grains (bullet not seated) meant that medium burn-rate powders would yield the best results within the bullet weight range I was using. Staying with my remit of 40 grains per case, I finally chose a 129 grain SST Hornady bullet and 40.0 grains of Vit N150 powder, which resulted in a velocity of 2850 fps – a good performance for an 18in barrel. If you feel a little adventurous and appreciate top-end performance in a very accurate and mild-mannered rifle, the Rapier range of cartridges might be just what you need.

30-47L

This is another of my wildcats based on a 6.5 × 47L case, but this time necked up to .30 calibre to achieve an efficient woodland and stalking round. For more information, please see the Introduction above.

.300 Broadsword

I like .30 calibre rifles, so I wanted a .308 bolt face case to complement the 30-47L. That meant I could use a standard bolt head in my RPA Quadlite test rifle, but I wanted a fat, short case that would give a short powder column yet still achieve a good powder capacity. I looked at the WSSM, 45 Blaser and several old English cases, but settled on the .458 SOCOM, which was originally designed to shoot a heavy 350 grain payload from a M16/M4 upper conversion at 1800 fps. Its water capacity of 61.61 grains in only 1.573in overall case length was ideal. I could neck it down to the required calibre and experimented as far as 0.20 calibre, but settled on .30 cal. In that calibre, with a 125 grain bullet loaded, the whole overall length of the 300 Broadsword is still only 2.310in, smaller than a .308 Win case but with a 56.0 grain maximum powder capacity.

The barrel was a Walther stainless steel profiled to heavy varmint and fluted to reduce weight. The rifling was 1 in 10in twist to stabilize a range of bullets and the overall barrel length was a long 28in. For initial testing it was bedded into a McMillan A5 stock as it made a solid and stable bed from which to test the loads.

Forming brass is where it becomes a bit tricky. Trying to neck a .458 calibre neck down to .308 calibre is a big step and needs numerous individual reductions in order to stop the cases collapsing in on themselves or pushing a shoulder back. You can have a custom set of dies made with replacement neck bushes, but I acquired a set of CH4D large bushing neck sizer dies, which allowed me to reduce the neck diameter slowly down in increments of 10 thou (and sometimes

The 300 Broadsword was created to shoot like a .30-06 but in a case smaller than a .308 Win.

The first stalk with the 300 Broadsword was a success, as shown by the roe buck seen here.

5 thou) until I had the desired .308 neck diameter: it took nineteen individual neck bushes, Imperial Sizing Wax and two-and-a-half hours for ten cases. The final bush size was 0.340in but, as you can imagine, all that brass bending had made the necks thicker. The inside neck diameter was now 0.306in – 34 thou of neck thickness or 17 thou per side – so I needed to remove 5 thou on each side with the Sinclair neck trimmer to allow good neck tension and good fit into the 0.339 thou neck on the chamber reamer. It is a slow but very satisfying process when done by hand, although you can use a slow drill to speed it up, if you wish. Next you need to trim the cases to a length of 1.60in and then de-burr inside and out. Just for good measure,you should also de-burr the primer hole and re-cut the primer pocket for exact fitment. It is a right fiddle, but that stubby little round loaded up with a 125 gain Ballistic Tip is worth it.

I was hoping for a cartridge no bigger than a .308 Winchester in length but producing nearly .30-06 ballistics with a powder charge the same as a .308. This sounded fanciful, but it does hold 56 grains of powder in a short, dumpy profile. Again QuickLOAD came to my aid and predicted some loads to get me started and fashioned some excellent trajectories: a 125 grain Nosler at 3156 fps with only 44 grains of Vit N133 powder and Federal Match primer, or giving 3244 fps with an absolute maximum of 46 grains. At 100 yards three shots consistently made 0.35in groups from the long Walther barrel. I could have stopped there, but I had to try a few more loads. A nice fox or crow load was the 110 gr V-Max with 48.0 grains of Vit N 133 powder for 3311 fps; this was superbly accurate with half-inch groups at 100 yards. The same bullet and 50.0 grains of slower Vit N135 powder gave 3338 fps and 2722 ft/lb, while Reloder RL15 with 52.0 grains gave 3393 fps. A 125 grain Ballistic Tip, another light bullet, gave good accuracy with 50.0 grains of Hodgdon Varget powder yielding 3175 fps and 2798 ft/lb. Stepping up the weight load to 150

The 300 Broadsword wildcat is formed from the .458 Socom case and requires some serious attention to reduce the neck from .458 cal to .30 cal.

grains, the Nosler Accubond bullet proved to be excellent: 46.5 grains of Vit N540 powder gave a 100 per cent powder burnt for 2892 fps and 2787 fps energy and again really small groups. Similarly 45.0 grains of Hodgdon H 4895 powder gave 2947 fps and 2893 ft/lb. I broke the 3000 fps barrier with 46.0 grains at 3045 fps and 0.75in groups. A subsonic load was 9.25 grains of Vit N32C, or Tinstar as it is known. With a magnum CCI 250 primer and Hornady 90 XTP bullet, this gave 1054 fps and was very quiet with a sound moderator fitted. As an alternative, a Lapua 200 grain B416 bullet with 11.75 grains Tinstar powder gave 1098 fps velocity. A Sabre Cut custom bullet saboted round weighing 48 grains can be pushed to 4457 fps with a payload of 50 grains of Alliant Reloder RL7 powder. It is a very efficient case design and accuracy was really good. Although the 300 TK, a wildcat based on the 300 WSSM case, is easier to form, it does not have the convenience of a 0.308 bolt face for easy and cheap rifle conversions. The name Broadsword will be remembered by all those familiar with the iconic 1968 film *Where Eagles Dare*.

500 Kimera

My current project, the 500 Kimera.

This is a wildcat designed to use a very British case as a dual-role super/subsonic cartridge. I wanted a case capable of propelling a .50 calibre expanding bullet at 1050 fps and meeting the deer legal minimum of 1700 ft/lb energy in England and Wales. It also needed to be able to launch a solid and expanding 600 grain bullet at 1776 fps for 4200 ft/lb energy for dangerous game. This brought me to the .425 Westley Richards case. It has a .308 Win case head, so will fit any rifle with this bolt arrangement and if you straight case the cartridge, by taking away the taper or bottle neck, it is a perfect fit for a 0.510 calibre bullet. True head spacing of the case is tricky, due to a straight-walled main body section, but the performance and dual-role nature of this round is very interesting and development continues.

The 500 Kimera was designed to satisfy my fascination with large calibre subsonic rounds as well as being a capable supersonic round.

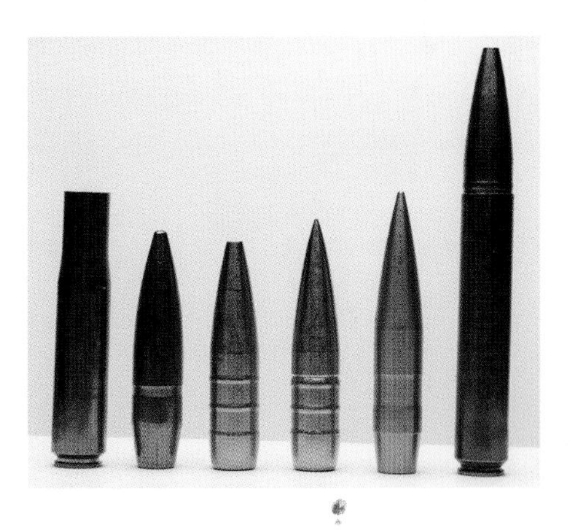

Based on the .425 Westley Richards case, the 500 Kimera was originally designed to shoot a custom-made 725 grain expanding bullet at 1050 fps (subsonic) for a deer legal 1800 ft/lb energy.

Chapter 7
All the Trimmings

Once you have determined all the major parts of your custom rifle, it is time for the part I really like: the extras. Again this is where an individual can express their personality as we all have our own ideas about the fine line between getting it right and going over the top. That's one of the main attractions of having a custom rifle. It's all those little details that actually make it personal.

ENGRAVING

Engraving is a very personal choice. When done correctly it can add to the overall appearance and desirability of any custom rifle. There are many engravers in Britain who excel at this craft, primarily for shotgun projects, although rifles look equally good with engraving to the action and sometimes to the barrel. Scope mounts are obvious areas where some engraving is attractive, but too much can overpower a good custom rifle's look.

Well-executed, simple designs such as fine scrollwork and acanthus leaves are still popular, as are areas of deep or heavy relief scrollwork. Some even choose game scenes of species the rifle will hunt. The skills needed to produce this type of work will contribute to making the rifle an individual thing of beauty that you will treasure and be proud to own and shoot. A popular area for engraving on rifles is the magazine floor plate, where a combination of scrollwork, game scenes and gold inlay always looks good due to the open flat surface and the contrast between the differing finishes. This is the first thing you see when the rifle is stored in a gun rack.

Laser-cut engraving has been increasingly used for cutting elaborate stock patterns. Some metal engraving by laser is also possible, but it never has the look or feel of hand-tooled engraving. If done well, gold and silver can embellish a rifle tastefully, but too much detracts from

A very stylish Anglo Custom Rifle Company stalking rifle.

Engraving really sets off the metalwork, as shown by the combination of classic scrollwork with animal relief on a Tikka 55 .243 magazine.

Classic scroll and lettering emphasizes a rifle's lines and model number.

the subtle lines of any true custom rifle. Fine scrolled engraving can be emphasized by a thin border or fine ribbon of gold. Makers' names or model type and numbers in gold stand out well against a blued background finish. I find that a single gold ribbon around the muzzle on a classic sporter adds a touch of class and picks out the fine detailing on either the magazine floor plate or scope mounts. If done well, figures or animals in gold inlay can also look good. As a general rule less is more, although some rifles can be totally gold, silver or brass, or specific areas such as the bolt or scope mounts can have a coated finish.

One method is to use a Forster gold inlay filling kit. After the surface has been cleaned,

Subtle areas of engraving lift the metalwork, as on this Norman Clark classic .280 Winchester custom sporter.

Pistol grip caps are well suited to engraving, as shown by the fine scroll-work, bordering and leaf detailing on this Jefferies. The spaces between help to accentuate the fine detail.

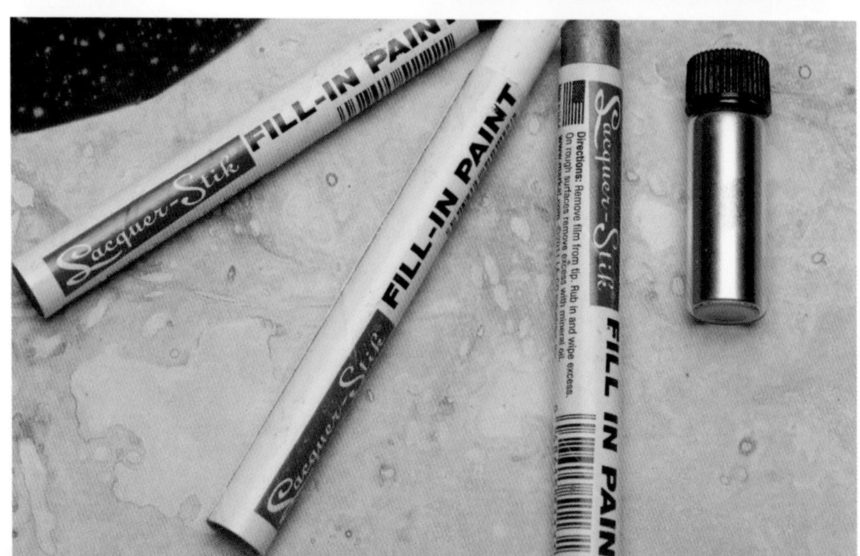

Gold, red or white accents can be made using gold paint or Forster paint sticks, which work well for lettering accents.

Fine gold-lined borders always look good, provide the effect is not overdone.

Gold animal figures are not to everyone's taste and need to be done very well if they are to look good.

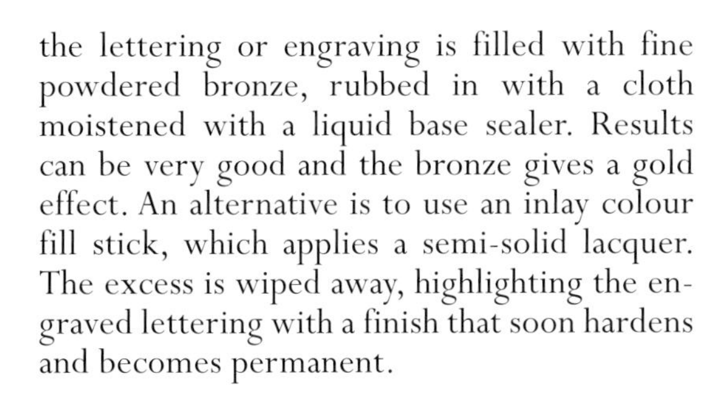

the lettering or engraving is filled with fine powdered bronze, rubbed in with a cloth moistened with a liquid base sealer. Results can be very good and the bronze gives a gold effect. An alternative is to use an inlay colour fill stick, which applies a semi-solid lacquer. The excess is wiped away, highlighting the engraved lettering with a finish that soon hardens and becomes permanent.

STOCK CARVING

Done correctly, stock carving can really enhance a rifle. Delicate adornments around the chequering and rear butt sections of a wood stock can be attractive, but some stocks that are nearly covered can look tacky.

The lovely walnut stock for my Tikka LSA 55 .308 Win rifle was sympathetically carved

Carving a stock is a fine art. This example of basket-weave chequering and acorn by Paul Richins adorns my Tikka 55 .308 rifle.

Paul Richins from Spirit Carvings hard at work on another masterpiece.

by hand with chisel and power tools by Paul Richins, whose business is called Spirit Carvings. Designs can be as personal or artistic as you like. The basket-weave and decorative oak leaf design in deep relief is superbly detailed and executed. A decorative oak leaf and acorn pattern in high relief was placed to the right and rear of the stock for balance. Carving as part of the chequering system or bordering can be effective. When it is used instead of chequering, for example with the basket-weave pattern on the Tikka stock, it has the further benefit of giving a very good grip. Some customers have requested carved animal designs on rifle, shotgun or air rifle stocks, or indeed anything wood. Paul has carved foxes, pheasants, rabbits, hares and rats for hunters, pest control firms, forestry workers and many others. The choice is yours.

COLOUR CASE HARDENING

Colour case hardening is my favourite embellishment. Truly traditional looking, it looks great on all manner of custom rifle projects. It is an age-old process that is still considered the epitome of quality and the finished result is a joy to behold, presenting a myriad of radiating colour variations that are unique and totally individual. No two finishes are the same.

The process of colour case hardening involves placing the chosen component into an airtight crucible surrounded by animal bone charcoal, wood charcoal and even charred leather. These are the carburizing agents in the process, which also hardens the outer surface of the metal (*see* Chapter 2). This is then heated to more than 1000°F (540°C) for between six and eight hours until the metal rifle parts are coated with the mixture. It is then placed over an aerated bath of water and the contents dropped in to quench the heat. It is important to aerate the water and not allow the metal surface to contact air before it hits the water, otherwise an oxide layer can form causing a flat, ordinary finish. The wild colour patterns are the result of the part being coated and the mixture cooling at different rates. This latter process takes place in milliseconds. The steam produced as the hot metal part touches the water reacts with calcium phosphate in the bone material to produce the coloured layers on the steel. Whole actions that have been colour case hardened look superb, especially on a classic Mauser-actioned rifle, but the process can also be used to subtly enhance areas such as scope mounts, bolt shrouds, pistol grip caps, air rifle compression chambers and moderators.

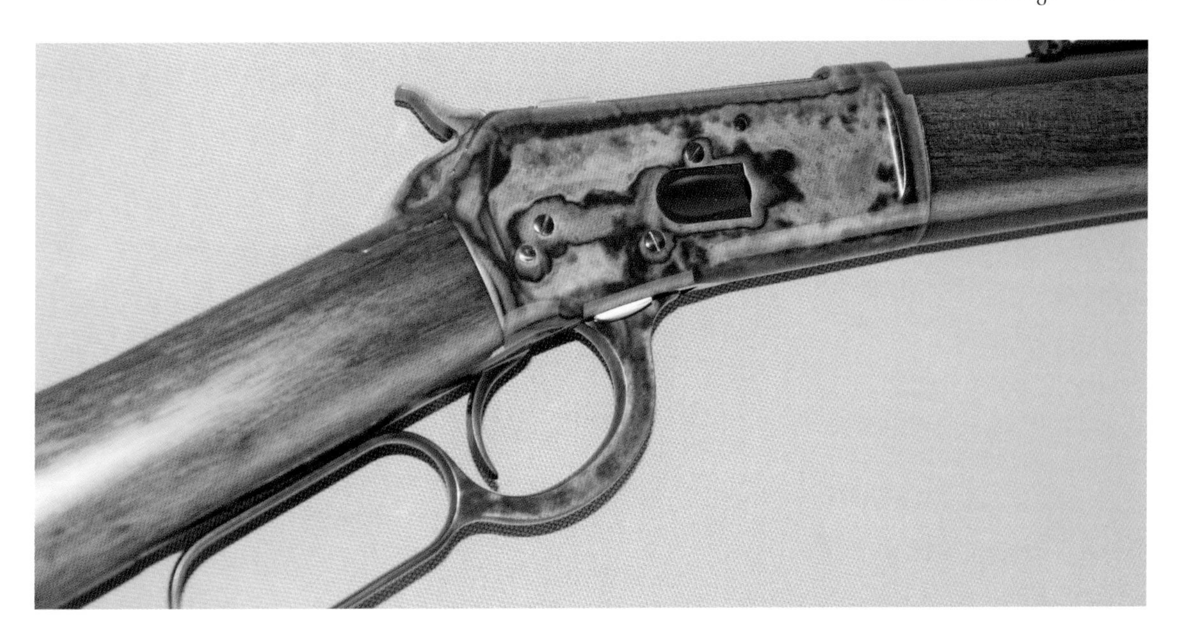

Colour case hardening is a long-established method of enhancing metalwork that still produces beautiful results, as on this Norman Clark Winchester action.

Case-hardened scope mounts always look good, but getting the colour and finish correct is very tricky.

Venom Arms Contender custom carbine with a case-hardened action by Ivan Hancock.

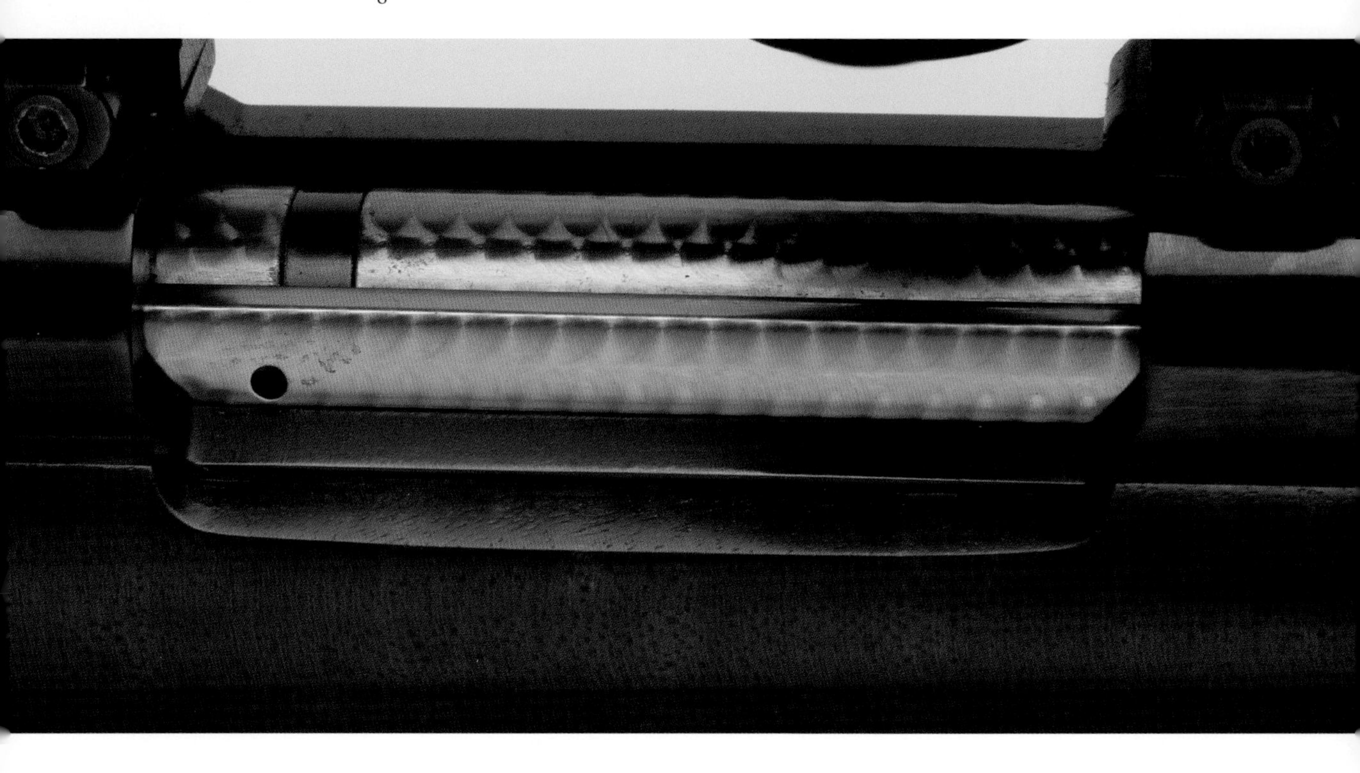

Bolt jewelling is another traditional method that enhances the bolt's appearance.

JEWELLING

Jewelling is the simple, if well-executed, process of turning a spiral pattern into a metal surface. This is usually most effective on a rifle's bolt, where it gives a three-dimensional effect without fluting and is cheaper. It serves no function other than ornamentation, but it can lift a solid piece of stainless or blued steel from the normal to the exciting. Best results come from a polished surface on a bolt or air rifle compression chamber or floor plate, with subsequent bluing to give a lasting shimmer and depth of appeal.

The process is not complicated, but it takes a true custom gunsmith to get it right and I have encountered some shocking jobs over the years. Venom Arms and V-Mach have been responsible for the best I have seen. The jewelling process uses an abrasive brush and compound that roughens the metal surface in a circular pattern. It can be applied to steel, stainless steel and some aluminium, if it is not too soft. Jewelling can also hide shallow scratches or imperfections that other finishes cannot.

MUZZLE BRAKES

These devices fit to the muzzle of the barrel via a thread section and redirect the muzzle blast in order to reduce recoil. This is a simple and cost-effective way to reduce heavy recoil on hard-hitting rifles like the .458 Lott. One thing to remember is that the exit ports or slits in the main body have to be precisely machined, otherwise the blast is redirected towards you.

A good muzzle brake reduces muzzle flip and recoil by directing the blast down and back, but not at you. It is not simply a process of drilling holes in a cylindrical body. True custom brakes have precisely radiused holes or slits and are works of art in their own right. They typically have radiused holes around the circumference of the cylinder, while others have ports and slits or a combination of the two. There has also been a trend to vent or pierce the front face to allow easement of the gases, leaving the bullet's path unimpeded as it flies through clean air. Some sound moderators employ the same principle. Many types,

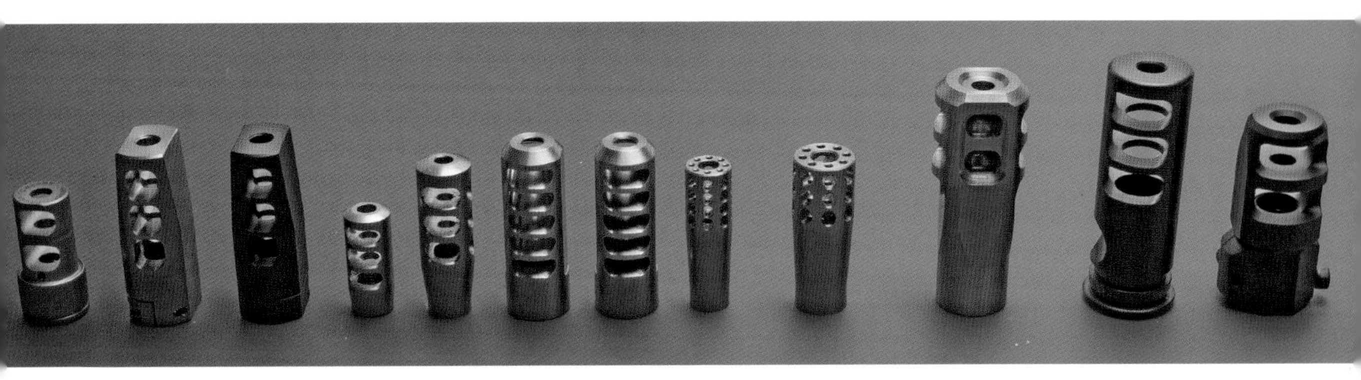

Muzzle brakes can be a necessity for hard-kicking guns or simply provide a good thread protector if a sound moderator is used. These muzzle brakes are designed and built by the Dolphin Gun Company.

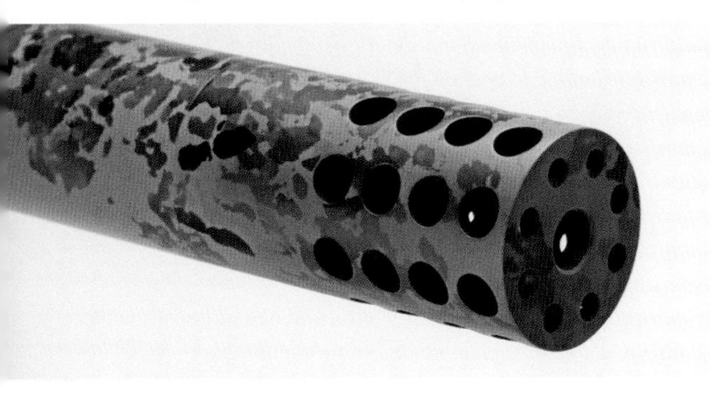

The pepperbox design of Valkyrie's excellent custom-built muzzle brake is quiet and highly efficient at reducing muzzle lift.

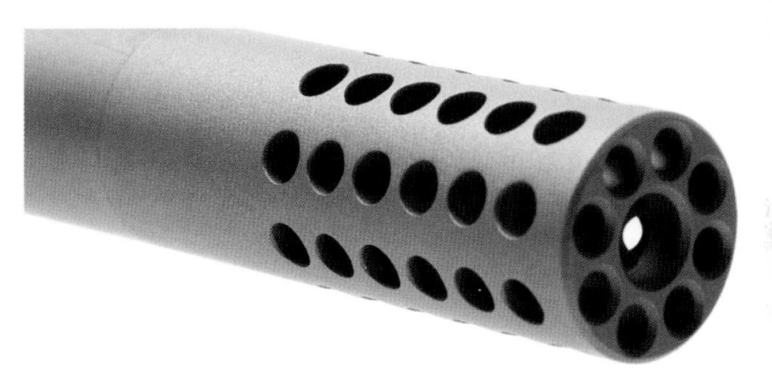

This custom .308 rifle shows that Riflecraft also knows how to build an efficient and quiet muzzle brake.

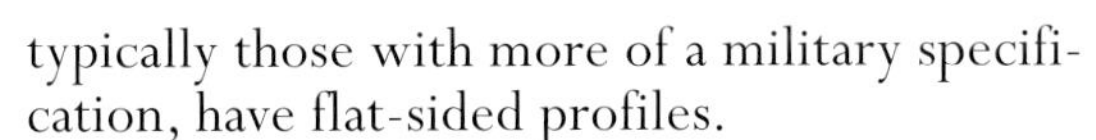

typically those with more of a military specification, have flat-sided profiles.

Custom gunsmiths in Britain will make a muzzle brake to suit your preferred barrel profile or style and at a performance that perfectly suits your calibre.

CARTRIDGE HINGED CONCEALED TRAPS

Here is a custom feature that is as beautiful as it is practical. It adds to the value and increases the appeal of any custom rifle. The age-old problem of storing additional rounds, other than in the magazine, has led custom rifle makers to use con-

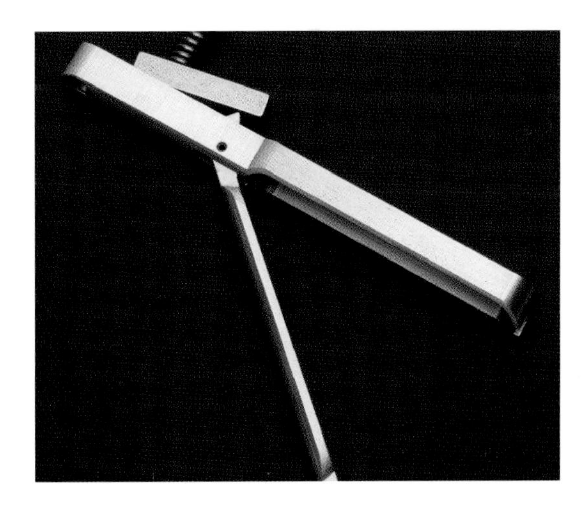

Classic stalking rifles always look good with a concealed cartridge box sited in the butt section of the stock. A Recknagel is here seen in the white before drilling and bluing.

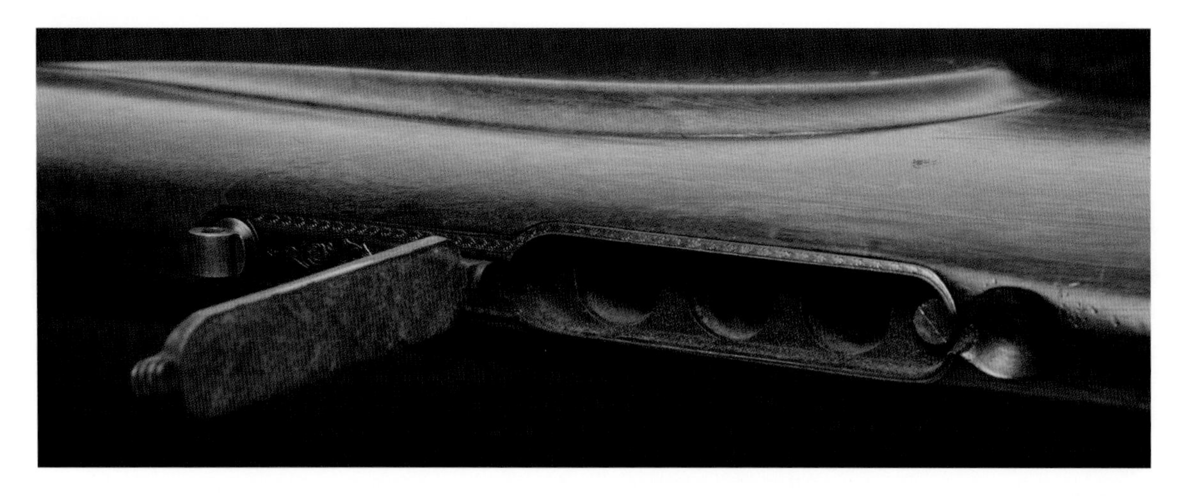

A cartridge trap can be useful as well as lifting the look of a custom rifle.

cealed compartments within the stock to store additional ammunition. As one would expect from any custom rifle project, this has become another feature to be decorated and engraved.

Cartridge traps are usually sited in the butt section of the rifle's stock. A custom smith is able to drill and mount an additional capacity of up to five rounds (in small calibre) to enhance the rifle's performance. The steel construction can be blued to any style required. A precise sprung lid is opened by a stepped latch to reveal the trap door. Most smiths will then finish the inside with a green baize or velvet, which holds the cartridges snugly so they do not rattle around.

Due to the continuing popularity of the Mauser 98 action, aftermarket custom parts are plentiful, such as the Recknagel custom bolt shroud, extractor claw and bolt release shown here. Specialized magazines are also available for large calibres.

MAUSER 98 CUSTOM PARTS

A whole industry has grown up to supply the custom market for the many classic custom rifles still built on Mauser 98 actions. Recknagel has a superb catalogue of precision custom-crafted accessories for the 98 action, including the replacement trigger units described in Chapter 5. For those building a traditional African rifle suited to shoot the classic old English calibres, such as .375 H&H or 404 Jeffery, Recknagel offers a magazine conversion to fit these long cartridges perfectly and feed faultlessly in the field; this is available in standard or Rigby-type styles. These are complemented by the precision-made extractor claw for standard or magnum calibres. This is a key part of any Mauser action due to the controlled round feed, and you need a positive cartridge manipulation in and out of the action when a Cape buffalo is breathing down your neck.

Another attractive item is the custom shroud for the bolt, which also is available in a selection of styles. All types have three-position safeties and are available in domed, flat and extra flat, shown here. The three positions are easy to use and the Secura option allows an additional inset lever to lock the safety in the third position until needed to be released and fired. All models have a gas shield in case there is a primer rupture.

STOCK CHEQUERING

Chequering is a common means of decorating a rifle, but when done well it has visual impact and a practical purpose in helping the grip. It is a traditional craft that requires a steady hand and keen eye. Patterns, coverage and cut orientation can really make a difference. Simple designs usually consist of a diamond or inverted diamond pattern on the pistol grip and forend, using a standard 20–22 lines per inch cut.

Pointed chequering, which has sharp points that improve the grip, has the advantage that it can be kept clean more easily than most chequering patterns. Another type is English flattop chequering, which can be seen on more classic rifles. It has very shallow and less sharply defined points that look good but can become clogged with dirt, hampering grip.

Chequering is usually measured by the number of lines per inch cut. Cutters are available to cut 16, 18, 20, 22, 24, 26, 28, 30 and 32 lines per inch (lpi). Cutters are available with single, double or multiple cutter blades for angle cutting from 60 to 90 degrees. For normal daily use a chequering pattern of 20 lpi is good enough, but custom rifle owners often like finer lines per inch as it looks very classy and is harder to execute accurately. You can enhance this by modifying the pattern to include a double diamond pattern or an elongated rear quadrant for more coverage, or add to this a double border or fleur-de-lis design. I personally like a wrap-around chequering to the forend, from where the chequering is continued around the bottom portion of the stock. It aids the grip but also looks good. The appearance can be further enhanced by chequering the top portions of the pistol grip where your thumb rests. These details may be subtle, but they can really stand out and show the rifle is not a stock item. Another idea would be to have different chequering on the inner areas. Instead of diamond cuts, you can mix up the effects with skip one chequering, where a line or cut is missed to give a dou-

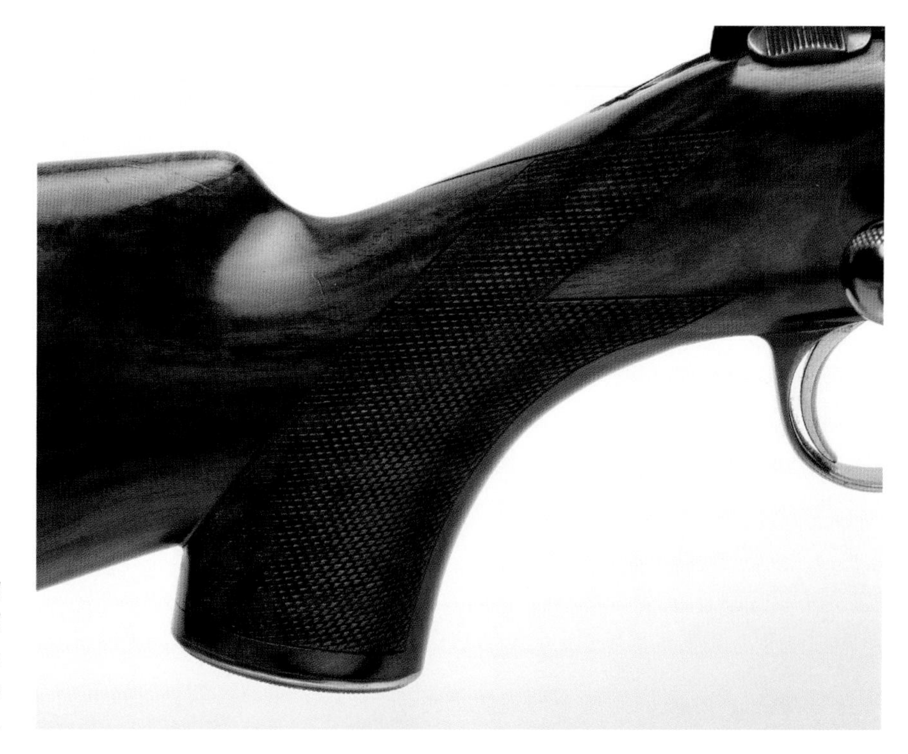

Well-executed standard chequering, shown here, can enhance a good wood stock.

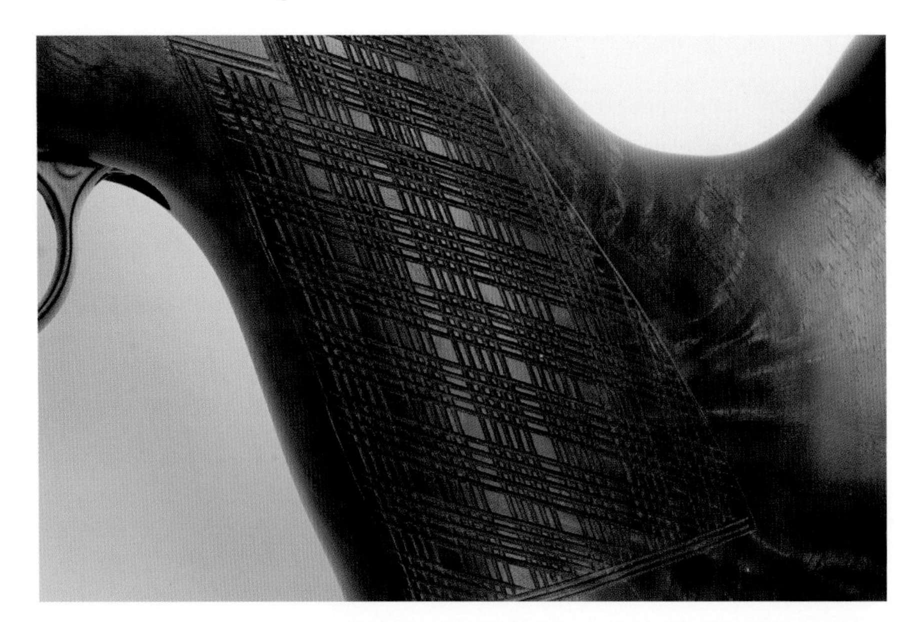

Skip-line chequering gives an alternative pattern that suits this Venom custom Weihrauch HW77.

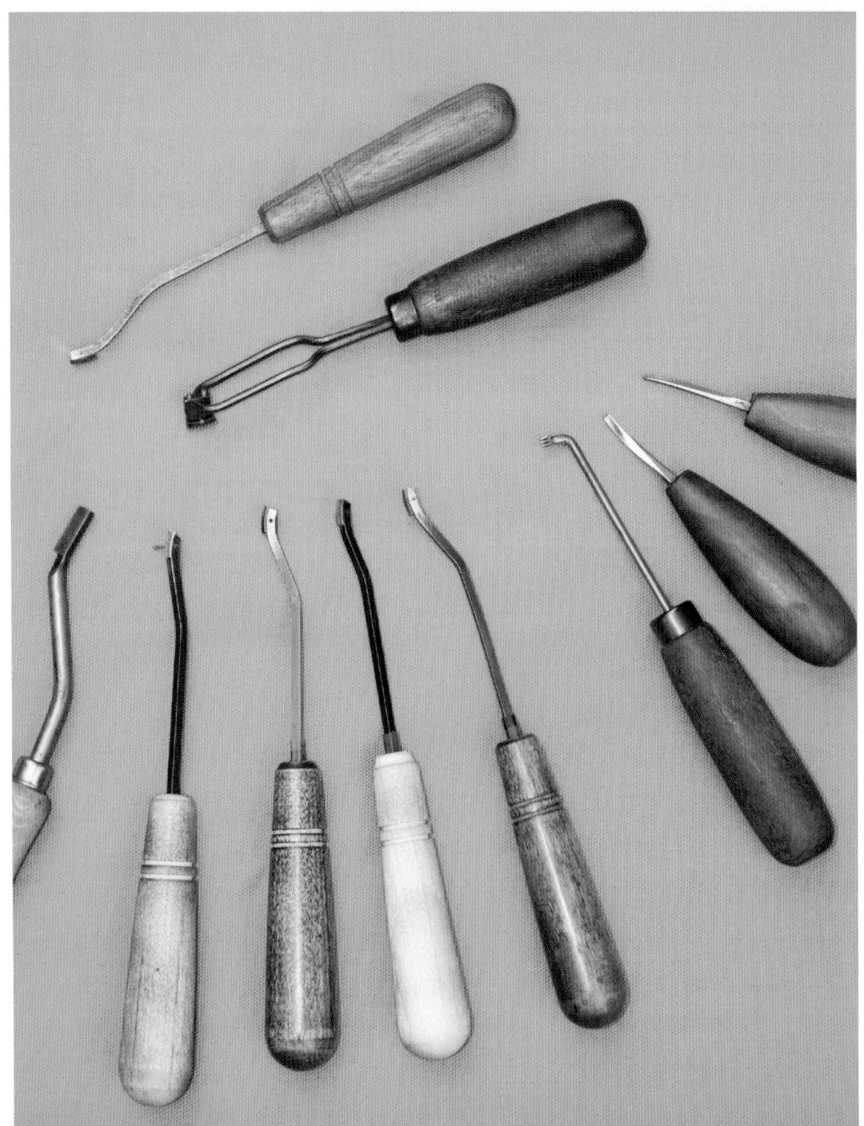

Rob Libbiter is a master at using a huge range of tools to produce precise, well-defined chequering patterns.

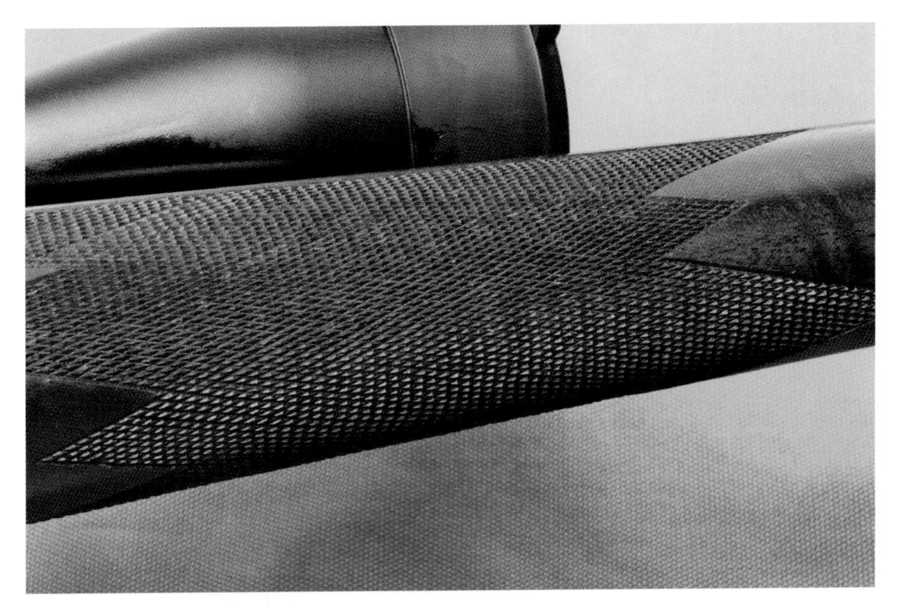

The wrap-around chequering to the forend of this Norman Clark custom Winchester M70 looks and feels superb.

Deep relief chequering on the walnut stock of an Anglo Custom Rifle Company Impala actioned rifle.

ble-sized diamond. This attractive style is sometimes called Scotch chequering.

Stepping up a gear, and more expensive, you can have basket-weave or fish-scale chequering, which requires a very skilful and steady hand. This type of chequering pattern is often combined with some bordered carving or extra lines, which really lifts the look. Deep relief chequering shows the true beauty of the craft. Deep recessed panels allow precise positive and true diamond-patterned chequering to emerge. The custom wrap-around feature is hard to execute properly but when done well, with a fleur-de-lis pattern intertwined within the main pattern, you have a custom chequering pattern worthy of any rifle.

Not all wood types will take chequering well. Some woods are too soft, so the finer you go, the worse the chequering looks. Laminates can be chequered, but they look better stippled. Many standard factory rifles now feature laser-cut chequering, but a good piece of hand-cut chequering stands out a mile and makes a custom project with fancy wood a joy to own. The chequering or stippling moulded into synthetic stocks is never really that well defined.

Stippling is a good alternative to chequering for synthetic stocks. These are some of the tools needed to producing that sort of design.

STIPPLING

Stippling is a lot simpler to apply than chequering and is certainly a better option on synthetic stocks. When done correctly it can look just right on a sporter or thumbhole designed stock configuration. A stippled area on a stock improves the grip to ensure a better shot, but it also presents a distinct contrast on an otherwise bland stock. A low-grade walnut stock that has been chequered will still look ordinary, but have it stippled and it looks like you have made an effort. That's the beauty of custom rifles in a nutshell: the smallest details make the biggest difference.

Stippling is a good choice for laminated stocks as chequering does not always take well to laminated stocks. Stippling is applied by a variety of tools. N. Clark makes his own custom tools for all stippling duties. Each tool produces a uniform raised textured area that is surprisingly easy to grip, which of course would be the real purpose of chequering any stock.

STOCK TIPS

The simplest form of stock adornment to any custom rifle is the addition of an alternative wood type to that used in the main stock. If you use a walnut stock, a rosewood tip is commonly used to accentuate the custom stock and add contrast. This may be further embellished with white spacers to segregate the differing wood types, although it would not be my choice. Classic rifles often utilize buffalo horn as a tip to finish off the wood, or it could be another exotic material like ebony, maple, horn, zebra wood or myrtle (this does not have to be wood). The pistol grip cap is a good area to adorn with exotic wood types, while engraving may be enhance hinged floor plates or cartouches.

Buffalo horn is traditionally used as a forend tip.

Ebony tips contrast well against the main wood in the stock.

Rosewood is also traditionally used to accent this area of a wood stock.

INLAYS

Sometimes less is more. Stock designs that are already flashy, such as the Harry Lawson Cochise, can actually take even more decoration. Another way to personalize a custom stock is to use inlays, perhaps exotic woods that contrast with the base wood. A thin white or double white line emphasizing the inlay works well. Simple designs like stars, crosses or Celtic symbols are best. Pewter designs and precious metals with more intricate designs, however, can be difficult to inlay accurately but give that personalized touch.

Many people like to personalize a new rifle with a shield or decorate the pistol grip end cap with their initials or an engraved animal cartouche. These can be steel, brushed aluminium, stainless, brass, bone, silver – or indeed any material that looks good. Recknagel produces a good array of ovals and shields that are precisely made with a secure attachment spur.

You can go one step further and add alternate wood inlays, as with this maple inlay on my .20BR Lawson Cochise stocked rifle.

People like to add their own initials to custom rifles and this collection of Recknagel shields is supplied in white so that they can be finished to customer requirements.

STOCK BUTT PADS

The choice of the recoil pad greatly influences both a rifle's looks and its function. Sporters, classic and lightweight mountain customs always look best with a slim butt pad in a solid colour. A brown, black or classic oxblood red, for example, sets off a good walnut stock. Many will go for a ventilated recoil pad of the same colour choice. This makes sense on a heavy-kicking rifle as they absorb recoil well but can look too large on some stock designs. This is an area where beauty really is in the eye of the beholder and a good custom stock can be enhanced or ruined by the wrong recoil pad.

Many new firms offer recall pads with a traditional outside appearance, but with technologically advanced honeycomb recoil-absorbing properties hidden inside. A classic-looking rifle with exceptional figured wood will always look well with a metal or skeletonized metal butt plate. These appear less cumbersome than rubber plates but require a good gunsmith to fit them correctly. Recknagel metal pads are supplied in the white or plain steel and can be blued, colour case hardened or engraved, or a combination of all three finishes can be used to enhance an often forgotten part of a rifle.

A solid red butt pad is the best option for any classic wood-stocked custom rifle.

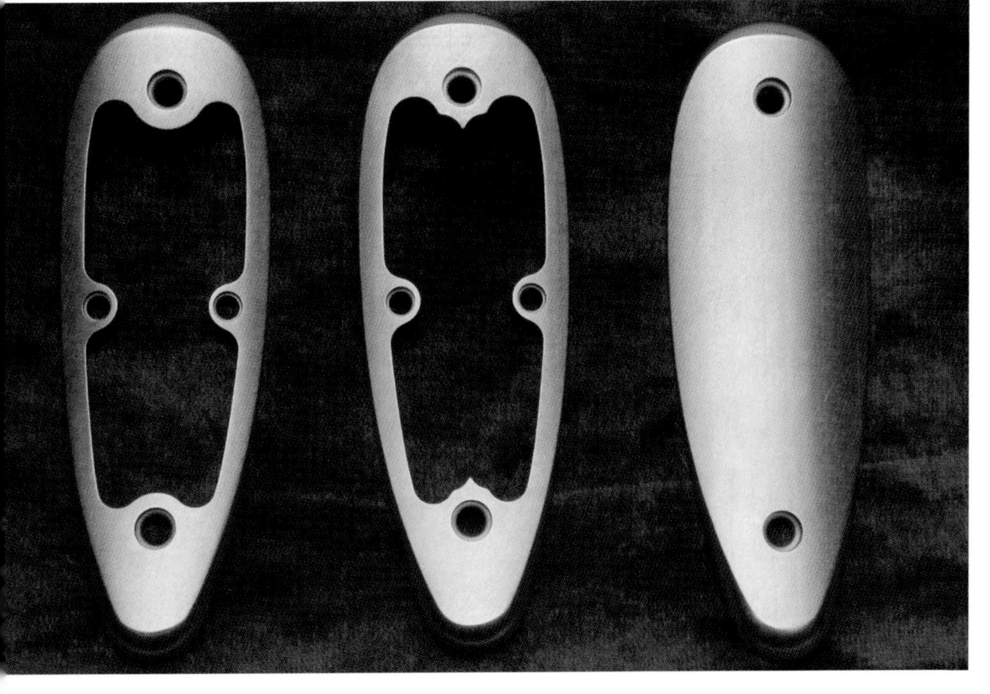

Recknagel's choice of traditional solid metal or skeletonized metal butt plates sets off a beautiful walnut stock.

Fully adjustable recoil pads, as on this stock design by the Dolphin Gun Company, offer the ultimate in custom fit and function.

BARREL BANDS

Barrel bands are very old-fashioned and traditional but this is a small item that can instantly make a big statement. Although they look best on a classic walnut-stocked rifle, they can lend an air of sophistication to any rifle. Some may say it interferes with the barrel's harmonics as it is there to support a sling swivel, but in practice it rarely has an effect on the first shot. It has some functional benefits in that it puts the point of balance and length at which a rifle is slung over the shoulder much lower, and so it is less likely to snag on trees or bushes as you hunt. Recknagel again offers a complete range of barrel band styles, sizes or fitments to help transform an interesting custom rifle into something special.

Barrel bands can be made to width and to support a sling swivel for sling fitment. They can be open-ended or completely rounded with a removable sling attachment. The most stylish are the soldered-on one-piece bands made to fit exactly the barrel's contours or diameter at the point it is to be fixed. These give very clean lines. On heavy recoiling guns this method takes the sling swivel stud off the forend wood, where your hand can get snagged under recoil, and places it on the barrel out of the way.

Barrel bands are practical and look good on a traditional Mauser-type actioned custom rifle.

Recknagel offers barrel bands of all styles and diameters to fit differing barrel profiles.

Sling swivel projections look awkward, so these Talley concealed studs, which fit flush to the wood, are a better option.

CONCEALED SLING SWIVELS

There must be a better way than to put an unsightly sling swivel stud into a beautiful walnut stock with gorgeous figuring. Fortunately, concealed sling swivel studs take many forms, ranging from completely invisible stud fixtures recessed into the stock to precision-made twin sling attachments that serve as both a secure fixture and a refined attachment. Talley Manufacturing in South Carolina makes a range of hidden sling swivel features: if ordered in a steel fixture they can be blued or colour case hardened to further enhance their appearance.

BOLT HANDLES

Although bolt handles are functional, this is another area that can be enhanced to benefit handling and add some individuality. Lengthening gives a better grip potential, as does increasing the size or configuration of the bolt knob it-

The bolt handles designed and made by the Dolphin Gun Company are both practical and add some flair.

self. Chequered, fluted, knurled or large plastic knobs all help provide a better grip.

BOTTOM METAL

Bottom metal is the term used for the floor plate or magazine assembly of the action. The traditionally floor plate design is fixed at one end and depressing a release button, usually sited on the trigger guard, allows the plate to hinge open so that cartridges can be removed quickly from the action. Because this usually has a flat face, custom makers are able to run riot with faceted, engraved or colour case hardened additions. Stainless steel floor plates from Sunny Hill are popular for stainless steel-actioned classic rifles. Detachable magazines, however, are becoming more popular. Magazine assemblies by HS Precision and Wyatt, and those compatible with rifles by Accuracy International, are the most popular and allow a higher capacity or just a quick magazine change when necessary.

A stainless steel Sunny Hill floor plate offers an alternative to alloy magazine floor plates for Remington clone actions.

Detachable magazines are especially popular on tactical custom rifles. Here is a Manners mini chassis bottom metal.

FINISHES

Bluing

Bluing has long been the traditional finish to the steel parts of custom guns, much like that of the oiled finish to the walnut stock. It gives a very appealing finish as well as some resistance to wear or to the weather. This type of bluing can be split into several processes and each custom smith will probably have their own secret recipe and procedure.

The first of these is called the hot blue process. It involves protecting the action, bolt, barrel or other steel parts of the rifle against

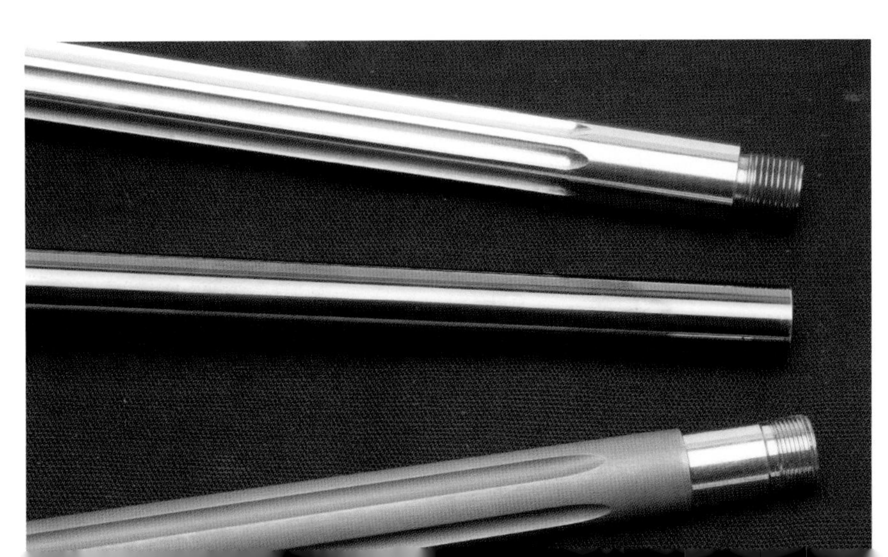

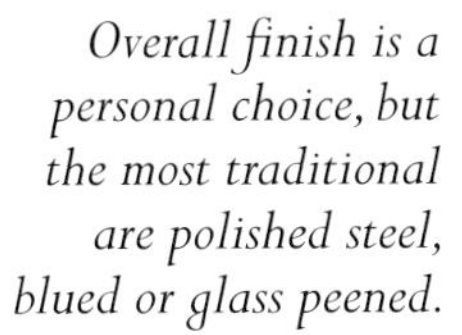

Overall finish is a personal choice, but the most traditional are polished steel, blued or glass peened.

rusting by immersing them in a tank containing a mixture of sodium hydroxide and potassium nitrate. This is then heated to above the boiling point. Timings vary according to the level of blue required and the steel used.

The cold blue process is when a paste containing the chemical selenium dioxide is applied to the bare metal without heat, using long, smooth and even strokes. This literally turns the steel black. It is then rinsed off and buffed up. Although it is tricky to get an even finish using this method, it is actually a good method for retouching threaded muzzles for a sound moderator or for small scratches that need to be re-blued.

Rust bluing and fume bluing are associated methods that are good for a traditional, classic look. They give a more subtle blue finish that can be varied depending on the number of times the process is done. Metal that is likely to rust is actually converted to black iron oxide (Fe_3O_4), which forms a skin of rust resistance and can be further enhanced by an oil treatment. The depth to the bluing can be maintained by the level and type of chemicals and timings used, but various effects can be made before immersing the metal parts in the bluing tanks. For a smooth traditional finish the parts are uniformly polished prior to dipping. A hunter-friendly matt finish is made by varying the surface texture of the metal with differing polishing pastes.

Stainless steel finishes usually have a dulled finish for hunting purposes. This is often called glass peening, where the steel is bombarded with aluminium oxide powder to give a dulled grey finish that is non-reflective, although it can mark easily. Although it reflects light, some prefer a bright polished action, especially for a benchrest gun and even some long-range varmint models, because but it does look very good. If you are after a concealed finish but want a stainless steel barrel, some form of paint or wrap is needed.

Paint

Most will baulk at the mention of paint, but a good custom paint job is actually very practical as well as attractive, if done correctly. Correct preparation is essential for a superior finish.

Rob Libbiter uses a spray booth to finish custom barrels, actions, stocks and parts with his military specification paint finish.

Rob's two-part paint finish is very hard wearing and a good choice for custom rifles that will earn their living out in the field.

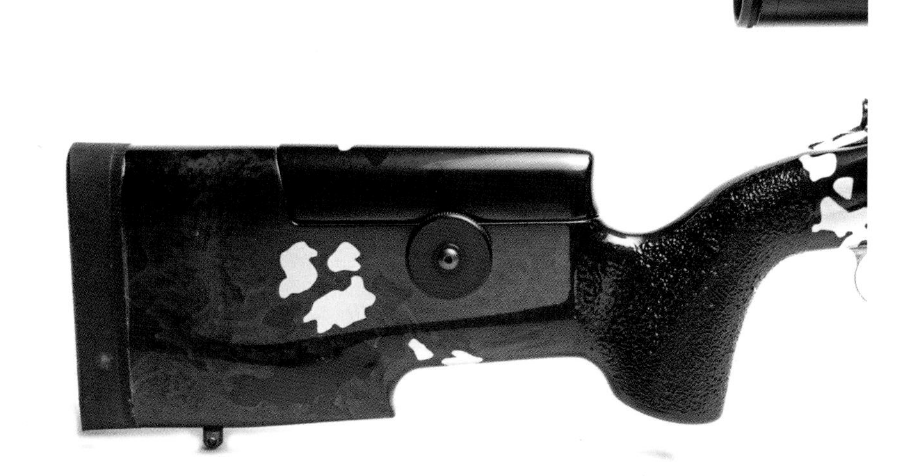

Gorgeous custom paintwork on a Valkyrie .22 BR rifle by David Wylde.

Custom paintwork like this Riflecraft all-over rifle coating is very effective and practical.

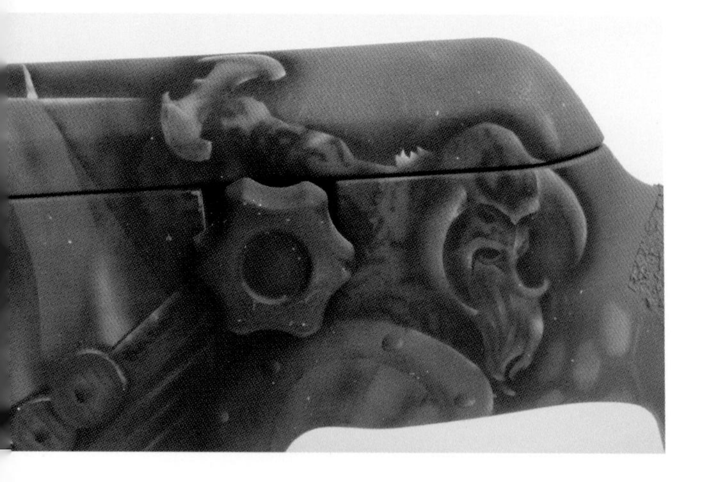

DCR custom murals can be applied to any custom rifle project. The only limit is your imagination.

Every flaw and pore must be sealed, especially around the bedding area and the magazine cut-out. The key is a good primed base coat that adheres to the wood, laminate or synthetic stock. The top coat can then be applied, followed by the clear lacquer.

Rob Libbiter uses a two-part paint system producing a military specification finish that protects the metal parts below, resists scratches out in the field and inhibits rusts. Many of my rifles are painted this way and it provides a tough exoskeleton that also has the great benefit of eliminating annoying shine and reflection from any stainless steel barrel. It is also available in an eye-watering array of colours and can be applied with stencils for patterns or as a custom pattern of two or more colours to suit your taste. You can choose to matching the finish of the mounts to the action for a fitted or blended look, or you can go the other way and make a statement by adding extra colour. A custom spray job to your design can really run riot, as these examples illustrate.

Custom Wraps

Unusual and striking wrap patterns offered by firms such as Prestige Custom Coatings will transform a plain stock. These are available in wood grain finishes as well as solid colours, marbles, crazy abstract designs and mural-type designs resembling those on custom motorbike tanks. An impressive design like one of these will make a statement and take your rifle to the next level of individuality.

A custom wrap instantly transforms the look of any rifle. Here is a Prestige Custom Coatings bird's-eye maple wrap on a standard beech stock.

Prestige Custom Coatings camouflage wrap is another option.

Cerakote

Cerakote is a ceramic-based finish that can be applied to metals, woods or synthetics to provide a weatherproof, stable, durable, impact-re-

A Cerakote finish is very tough. This Valkyrie finished rifle has twenty colour combinations for a unique and striking appearance.

Riflecraft's custom Cerakote finish can be applied in many alternate patterns to give an individual finish to any custom rifle.

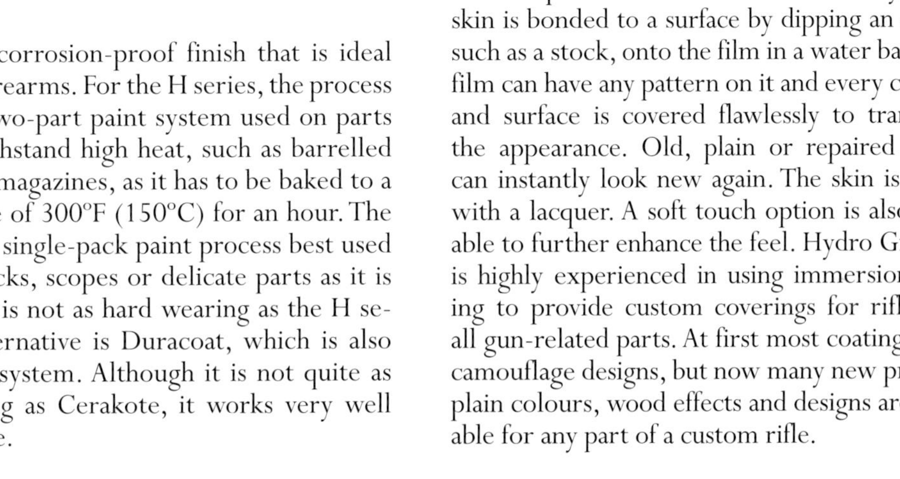

sistant and corrosion-proof finish that is ideal for use on firearms. For the H series, the process requires a two-part paint system used on parts that can withstand high heat, such as barrelled actions and magazines, as it has to be baked to a temperature of 300°F (150°C) for an hour. The C series is a single-pack paint process best used for rifle stocks, scopes or delicate parts as it is air dried. It is not as hard wearing as the H series. An alternative is Duracoat, which is also a two-pack system. Although it is not quite as hard wearing as Cerakote, it works very well for stock use.

Immersion Coating

This is a process in which a thin film or synthetic skin is bonded to a surface by dipping an object, such as a stock, onto the film in a water bath. The film can have any pattern on it and every contour and surface is covered flawlessly to transform the appearance. Old, plain or repaired stocks can instantly look new again. The skin is sealed with a lacquer. A soft touch option is also available to further enhance the feel. Hydro Graphics is highly experienced in using immersion coating to provide custom coverings for rifles and all gun-related parts. At first most coatings were camouflage designs, but now many new prints of plain colours, wood effects and designs are available for any part of a custom rifle.

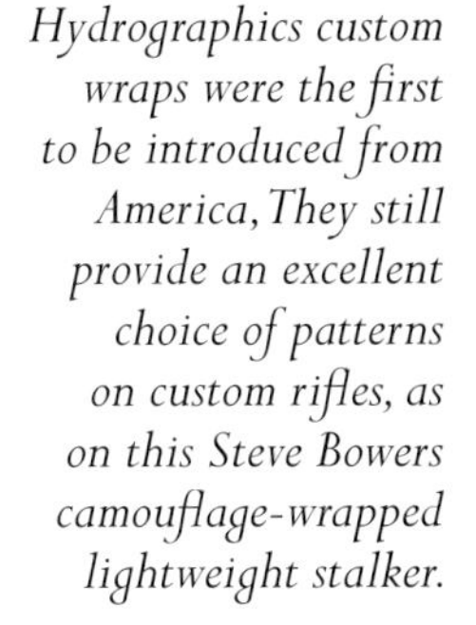

Hydrographics custom wraps were the first to be introduced from America, They still provide an excellent choice of patterns on custom rifles, as on this Steve Bowers camouflage-wrapped lightweight stalker.

Chapter 8
Custom Builders

Britain is well known for traditional custom rifle builders such as Holland and Holland, Rigby, Westley Richards and William Evans, but this book is concentrating on a selection of the new breed of custom builders. Many of these have responded to demand for a product that was either difficult to source from overseas or simply not available in the format desired by the customer. They include full-time gunsmiths, precision engineers who have turned their hobbies into a business, and talented individuals who have specialized in one area of custom work. The following discussions are meant to give a brief insight into their work and specialities, but in most cases the pictures will do the talking.

SPECIALIST RIFLE SERVICES

www.specialistrifleservices.co.uk
01242 863005

Specialist Rifle Services was established by Steve Bowers to build custom, high-precision, long-range, F-Class, varmint and stalking rifles. Steve is a fully qualified and experienced toolmaker who has been running a successful precision engineering workshop for more than twenty-five years. He applies his knowledge and expertise to producing extremely accurate custom rifles for shooters requiring a hand-built rifle for competitions, vermin control and deer management. Steve's interest in the countryside

Steve Bowers and his .30-06 AK improved chambered, barrel block bedding rifle, all of which was designed and made by Steve. It epitomizes the custom rifle-making skills available in Britain.

.20 Tactical on an RPA Quadlite action, McMillan A3 stock and custom-built MAE sound moderator.

BAT action chambered for the .280 AK Improved cartridge and Walther match-grade, fluted stainless steel barrel (blacked) and custom-built thumbhole laminated stock.

6.5 × 284 BAT HR action pillar bedded into a grey/black Bowers custom-built laminate stock and a Walther 1 in 9 rifling twist stainless steel barrel. It is custom-fluted and blacked by Bowers.

6.5 × 284 calibre chambered Krieger cut rifled barrel with a 32in barrel length and 1 in 8.5in rifling twist. The action is a Stolle Teddy with extended scope shroud and right bolt / left port orientation and CD-Universal trigger.

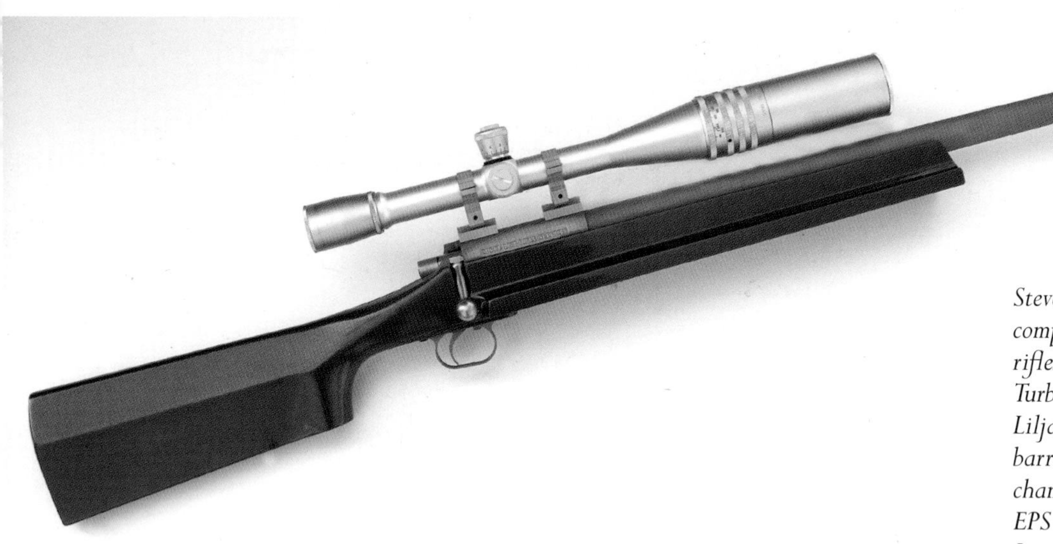

Steve makes precision competition .22 rimfire rifles, too, such as this Turbo V-1 action and Lilja 1 in 16in twist barrel of 20in and chambered for Eley EPS ammunition. A Jewell trigger is fitted and all is bedded into a benchrest style stock design and coated with a custom metallic paint.

This .22 Satan switch barrel RPA Quadlite actioned rifle was made for me by Steve to test my wildcat calibres. It is one of my favourite rifles for long-range vermin shooting.

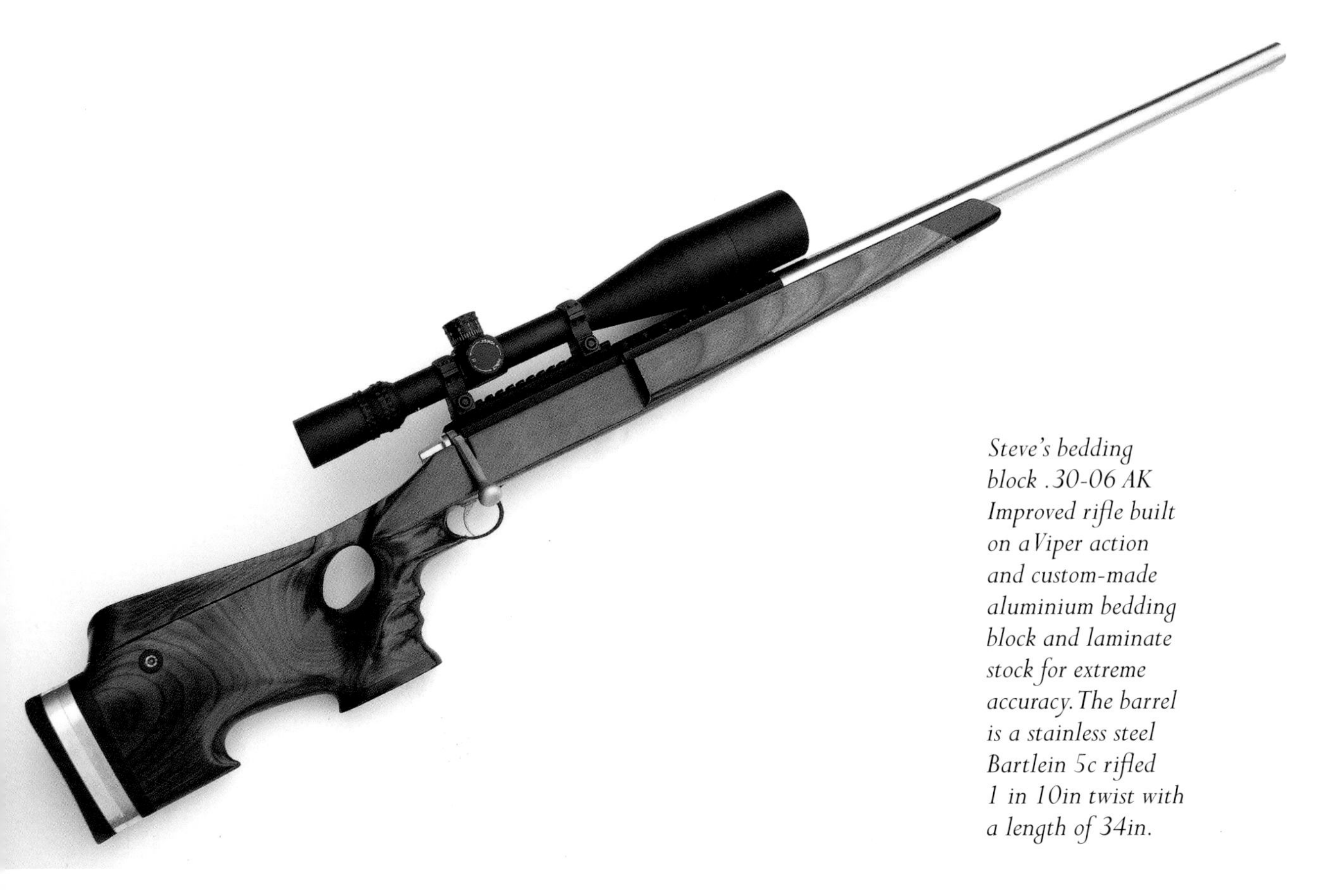

Steve's bedding block .30-06 AK Improved rifle built on a Viper action and custom-made aluminium bedding block and laminate stock for extreme accuracy. The barrel is a stainless steel Bartlein 5c rifled 1 in 10in twist with a length of 34in.

and shooting started when growing up in rural Gloucestershire and helping out with local pest control for pocket money. He joined an aircraft engineering company and later moved to an engineering firm to hone his skills. Operating from his comprehensive workshops, together with his in-house stock expert Rob Libbiter, Steve can turn any wood blank into something beautiful, as well as bedding, chequering and applying custom finishes. Steve has carried out most of my custom and wildcat work since Venom Arms closed. I can design the cartridges on QuickDESIGN and QuickLOAD, make the brass and develop the load, but I am not a gunsmith. Steve can turn my creations into reality and it is always good to work with someone who shares the same ethos and humour. He is just as capable re-barrelling a customer's shot-out rifles as he is building a custom rifle from scratch or blueprinting actions. He also designs his own actions, barrel-bedding blocks, scope mounts and any manner of one-off custom parts. That is where the precision engineering side of Steve's skill comes to the fore: no job is too difficult and his work is highly regarded and much sought after.

PRECISION RIFLE SERVICES

www.precisionrifles.com
01807 580422
The mention of Callum Ferguson's name will always receive a positive response in shooting circles. He is one of Britain's most highly respected and original custom rifle makers, building rifles to customers' specification for consistent accuracy in hunting or competition. Callum started his shooting life as a gamekeeper, later progressing to wildlife manager for a forestry company in the Borders of Scotland. His use of centrefire rifles for his job stimulated a desire to have a better rifle and so he joined Border Barrels Ltd, based

Borden stainless steel Alpine action with a Krieger cut rifled barrel and polished finish, chambered in 6mm-284 and pillar bedded into a McMillan Sako varmint stock.

Borden actions are superb and Callum's attention to detail is second to none. This example is also fitted with a Jewell trigger and Sunny Hill floor plate.

Callum's signature PRS custom rifle with McMillan stock fully bedded with pillars and a Borden Timberline action and Krieger cut rifled barrel, chambered in a calibre of your choice.

RIGHT: *Callum re-barrels many customers' own rifles, such as my Tikka M65 action with Shilen barrel in 7 × 57mm AK Improved and Lazzeroni thumbhole stock.*

RIGHT: *Callum also offers a bespoke wood stocking service, as seen on this custom Dakota actioned rifle crafted into a beautiful piece of walnut.*

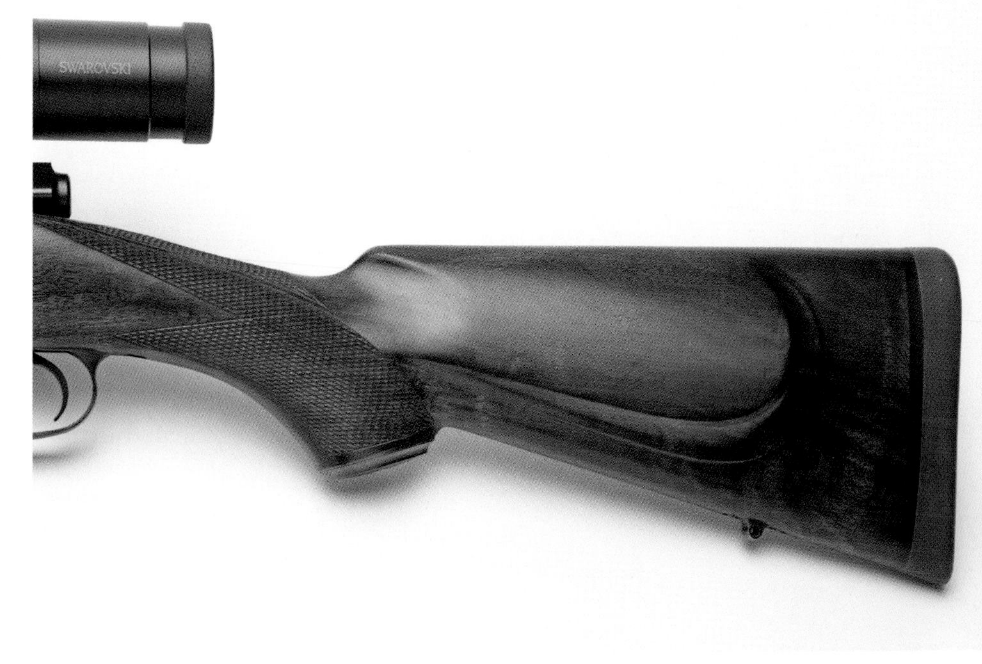

BELOW: *Beautifully executed lines, chequering and wood quality make a very practical and handsome traditional classic custom rifle. Here a Dakota 76 action in .25-06 Rem, with a 1 in 10in rifling twist rate and 23in length, and invisible screw-cut for a sound moderator, has been bedded into this custom-made walnut stock.*

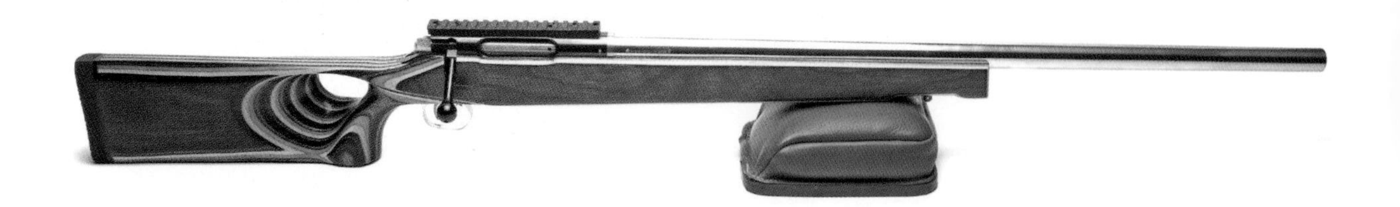

F Class FTR rifle built on a Barnard action and pillar bedded with Marine-Tex epoxy compound into a thumbhole laminate stock of striking purple, yellow and orange colour. The barrel is a Krieger 30in length chambered for the .308 Win round with a tight 0.342in neck and 1 in 12in rifling twist.

in Newcastleton, to learn the process of match grade barrel production. In 1990 he founded Precision Rifle Services with his wife Yvonne, focusing on building benchrest-quality custom rifles for all sporting needs. His rifles are works of art and capable of extreme and consistent accuracy, shot after shot.

Callum offers bespoke and personal service of the highest quality from his workshop at Strathavon Lodge, an old shooting lodge in Banffshire, which is always well stocked so clients can peruse the available components and then talk through the finer points of fit, finish and desired performance. He is able to produce a tack driving rifle in a traditional walnut-stocked guise or with a contemporary fibreglass-based synthetic stock.

NORMAN CLARK GUNSMITHS

www.normanclarkgunsmith.com
01788 579651

Norman Clark is a custom rifle builder of the old school, producing all manner of firearms from classic wood-stocked rifles to competition-winning F Class rifles. He developed his skills as a trainee gunsmith in the Birmingham gun trade under the tutelage of former W.W. Greener and Webley & Scott gunsmiths. He set up his own workshop in 1984, initially in a single rented room in Rugby. In time his business grew into a custom shop and retail business selling all kinds of gun-related products. As well as his custom rifles, Norman provides precision reloading equip-

.458 Lott chambered Krieger cut rifled barrel on a Weatherby modified action pillar bedded to a Lazzeroni thumbhole McMillan stock. The Krieger barrel is chrome molybdenum and finished with a superb blued finish and invisible thread protector.

.280 Rem rifle based on a Winchester M70 with a top quality blued Krieger chrome moly barrel, engraved action and handmade Claro walnut classic-style stock finished off with superbly executed hand-cut chequering.

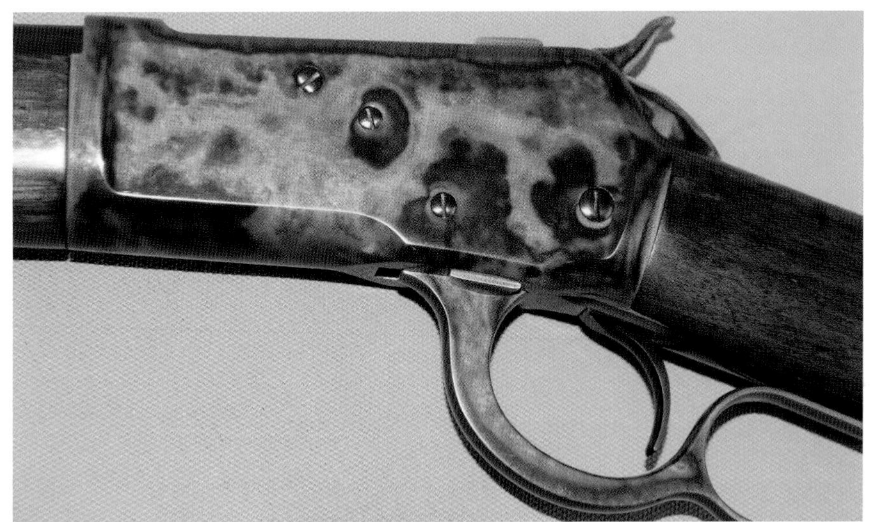

This restored, colour case hardened Winchester lever action has outstanding fit and finish.

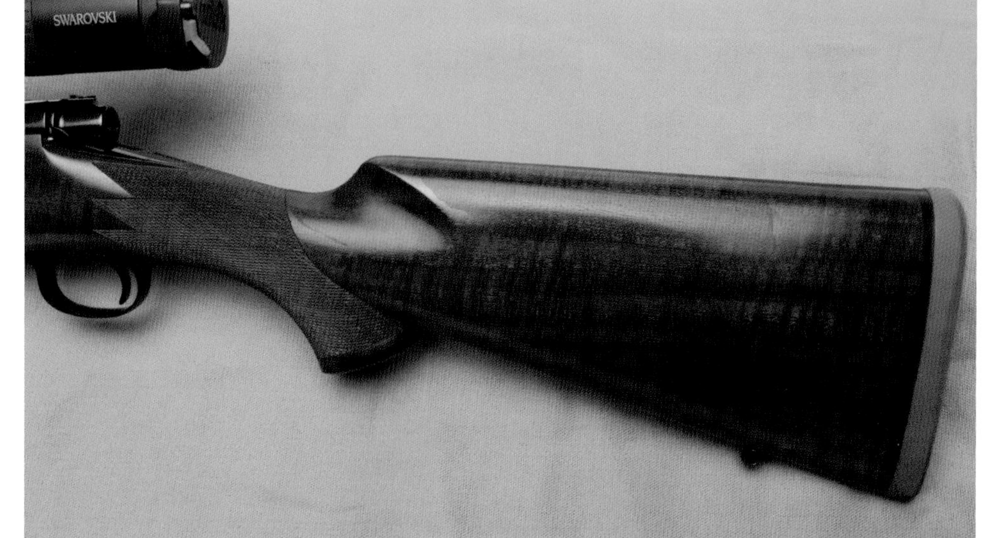

The classic flowing lines to the stock on this .280 Win custom, with excellent chequering, are set off nicely by the red recoil pad.

ABOVE: *Krieger barrelled Sako sporter profile barrel with magnum recoil lug fitted and custom high-visibility open sights.*

My .35 Whelen AK Improved rifle built by Norman on a Tikka action, bedded with Acraglas and a recoil arrestor fitted in a McMillan Sako-style stock and a Shilen chrome moly barrel.

ment and components to feed them with and has a wealth of knowledge on tap.

Over the years Norman has built and customized a wide range of precision stalking, varmint, tactical and target rifles in a variety of calibres ranging from .14 up to .505 Gibbs. If it burns powder, Norman can make it. Rifles he has re-barrelled, re-stocked and custom fitted for me include K98 snipers, a .35 Whelen AK Improved and a custom .458 Lott.

BROCK AND NORRIS

www.brockandnorris.co.uk
0845 521 2995

Mike Norris is well known on the custom rifle circuit. He is based in Sandford, Shropshire, where he started a custom rifle business in partnership with Steve Brock. He builds accurate rifles that real shooters want to shoot and live up to the slogan 'Rifles born in the field'. They are built as series of custom rifles, including the Ratal, Predator and Fulcrum, and each is custom-made from scratch to the customer's specification and fit. You can order a full custom, have a rifle re-barrelled or custom bedded, or just ask for advice as Mike has an almost encyclopaedic

ABOVE: *The Ratal or Honey Badger rifle is a Mike Norris signature rifle. It uses a short 19in tight-bored .308 Win barrel of heavy profile, all nestled into a McMillan M40AL HTC stock and finished in an Olive Drab Green Cerakote application.*

RIGHT: *The action is a Ratal one-off, designed in England for Mike. It has a Rem 700 inlet profile, but the whole stainless action is made to benchrest precision with a Tier One bottom metal and AI magazine.*

BELOW: *The Contractor is a very accurate but cost-effective rifle. It is short in stature but uses a Howa action and synthetic stock, and may be fitted with a custom barrel to give perfect handling and accuracy in the field.*

ABOVE: *The Fulcrum rifle has a Stiller TAC 300 action bedded in a McMillan Remington Hunter profile stock, like the Sako Model 75 stock design. The barrel is a Krieger varmint profile, beautifully fluted and of match grade quality, and threaded at the muzzle for a muzzle brake.*

ABOVE: *Each Fulcrum rifle comes with hard case, cleaning kit, loaded ammunition, test target card and load data.*

BELOW: *A Tikka M55 rifle that Mike rebarrelled for the 7.92 × 33mm Kurz round from the Second World War MP44 assault rifles. It is very accurate, quiet and a highly efficient for game the size of a roe deer.*

memory for reloading data and downrange performance. Mike does not just make a rifle and leave you to it after the bill is paid: each rifle is presented as a package with a recommendation to scope, mount and sound moderator types. Best of all, each of his custom rifles is range tested as proof of performance, with accurate load data and downrange trajectories provided with each rifle. Some models come with hard cases, cleaning kit and a quantity of precision loaded ammunition to get you started. Mike built me a re-barrelled and bedded Tikka M55 in 7.92 × 33mm Kurz round, the original Second World War German MP44 round. This is a rarity but superbly accurate in super- or subsonic guise and it just makes me smile.

DANE & CO

www.customrifle.co.uk
01892 864676

Paddy Dane started by building custom guitars, but he has become one of the new breed who did not see what he wanted on the custom rifle market and decided to put his engineering and artisan skills to good use and design his own. Paddy uses the latest developments in CNC lathes and milling machines in his well-stocked workshop. As each is a one-off custom special, he is able to re-barrel a rifle to benchrest quality in most rifle styles and calibres, including tactically stocked customs, heavy varmint long-rangers, F Class rifles and the usual assortment of sporter class hunting rifles. Paddy also shares my passion for odd calibres and shooters often request such cartridges as the .20 Dasher, 6.5 SAUM or .260 Nosler. Attention to detail is excellent, and build quality and finish are second to none.

An unusual 6.5mm SAUM chambered Benchmark carbon fibre barrel with a 26in length, 1 in 8in twist, 0.2962in neck diameter. This is bedded into a Manners MCST4A mini chassis stock.

Borden Alpine action with HVRTS Jewell trigger on the 6.5 SAUM rifle.

A 7mm-08 Border cut 5R barrel with a 1 in 9in rifling twist rate in light varmint profile of 21in overall length and a Lazzeroni thumbhole stock pillar bedded with Devcon.

The action of the 7mm-08 rifle is a proprietary branded Borden Alpine with a Jewell trigger and Sunny Hill stainless steel floor plate, with a Tier One 20 MOA scope rail and Spuhr SR3000 rings and a March scope.

A .17 Rem with 1 in 9in twist rate of 26in length and Pac-Nor Rem varmint-profiled barrel fitted to a Borden Alpine action in an AI AX2014 folding PSR rear-end stock.

Paddy fitted a custom-built MAE T12 sound moderator with calibre-specific baffle stack in stainless steel for highly effective noise reduction.

V-MACH CUSTOM RIFLES LIMITED

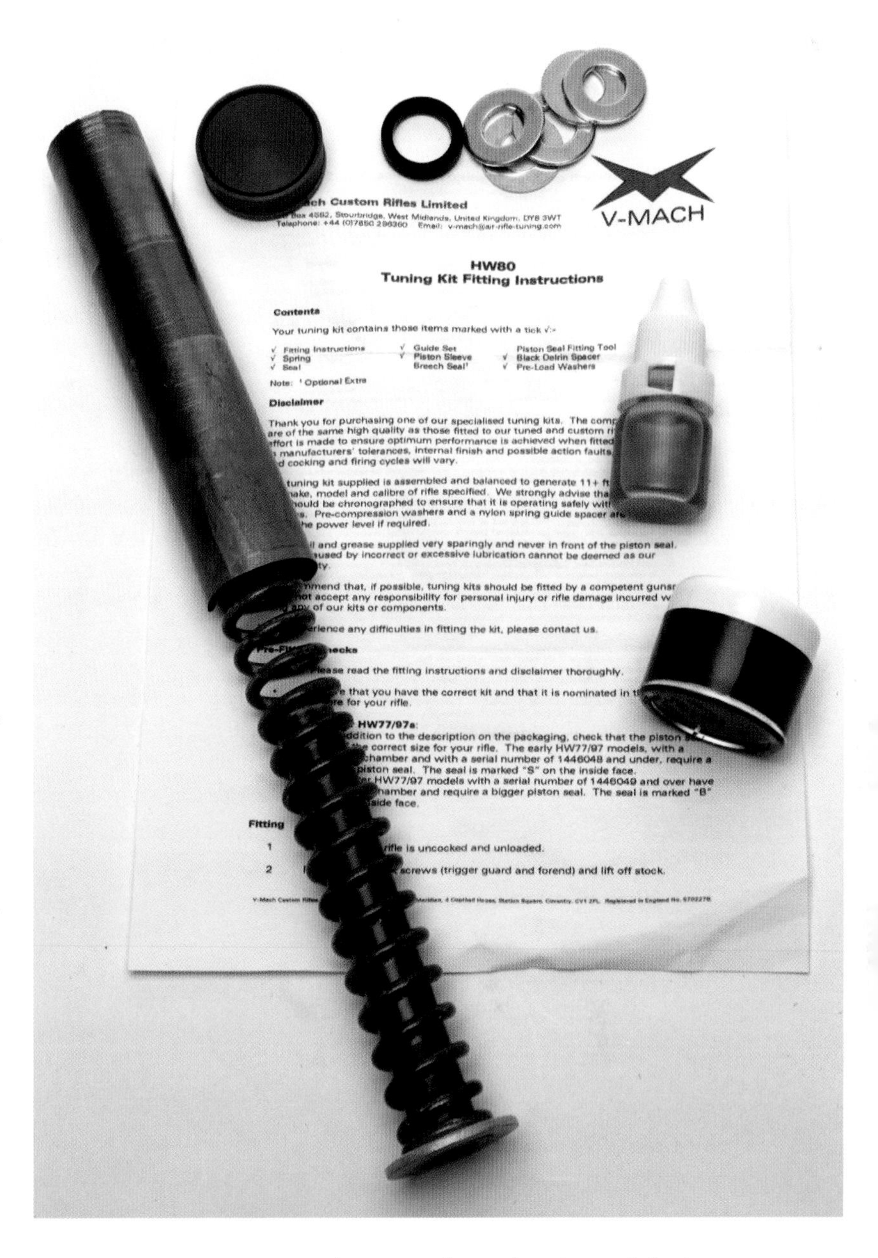

Steve Pope, the owner of V-Mach, is Britain's leading air rifle tuner. He can provide a kit for users to upgrade their own rifles, shown here, or can fit the V-Glide system to an air rifle in-house for ultimate performance.

www.air-rifle-tuning.com
07850 296360
Steve Pope is the son of Dave Pope, co-owner with Ivan Hancock of the famous Venom Arms Company. Steve continues in his father's footsteps as V-Mach, which was founded in 2006, and produces arguably the best custom air rifles in England. He is able to customize any air rifle made through his extensive knowledge, from a simple re-spring to a full custom gun. Steve started in a local gun shop, Trapshot of Lye, where he repaired and tuned customers' air rifles. From 1982 Venom Arms Company was established as a separate entity. This was the golden age of air rifle customizing, drawing on a very English style that no one else could replicate.

Field Target shooting was then in its infancy. Venom played a key role in co-founding one of the earliest Field Target air rifle clubs and kick-started the airgun customizing business in England, establishing its premium status. Venom Arms Company was responsible for some of the most innovative developments in the field, such as the Venom-Mach 1 and Venom-Mach 2, which remain the pinnacle of custom air rifle art – often copied but never equalled. The Laza-glide conversion and Steve's new V-Glide system remain the best possible custom upgrade options for any air rifle. The whole internals of any gun are re-honed, modified and then custom engineered to run on synthetic bearings to produce the most consistent and accurate spring air rifles today.

This old Venom HW77, which Steve's father co-owned with Ivan Hancock, has been tuned to perfection by V-Mach.

This .20 calibre Tomahawk custom has a totally new internal V-Glide system and custom thumbhole stock. It can be used to shoot one-hole groups at 30 yards all day.

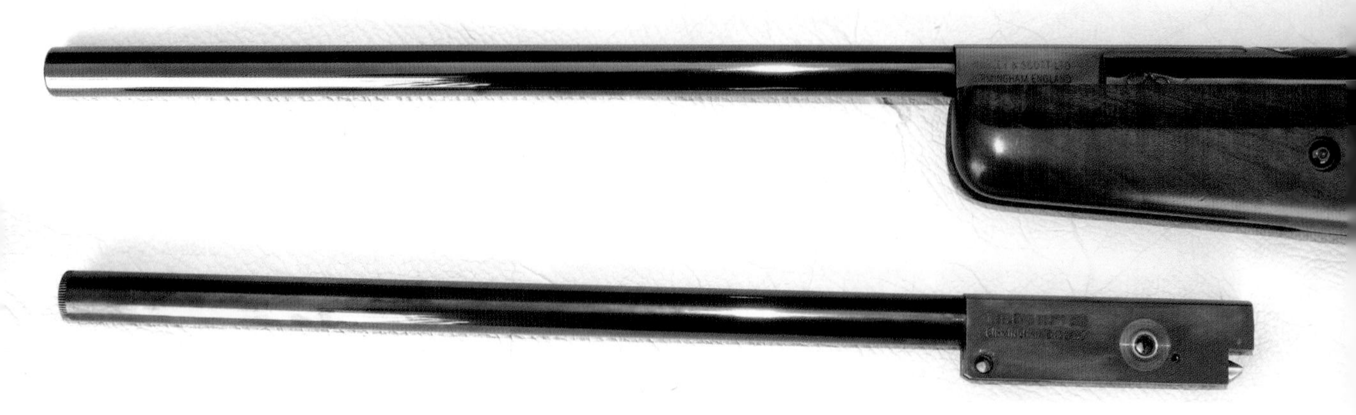

Steve makes many custom air rifle parts, especially sound moderators. Here is a switch barrel .22/.20 cal fully silenced barrel exchange system.

SWIFT PRECISION RIFLES

www.swiftprecisionrifles.co.uk
07739 689 871
After the closure of the Webley Venom Custom Shop, Richard Pope started his own business as Swift Precision Rifles. Although he has an air rifle heritage, Richard is best known for his custom full-bore rifles. He can re-barrel, re-stock, convert, bed, thread or build a full custom rifle as the client requires. He is a talented precision engineer who can turn his hand to any custom work from one-offs to carbon fibre sound moderators.

.250 AK Improved chambered Savage rifle with Pac-Nor super match-grade stainless steel barrel and custom bolt handle, trigger guard and Hydrographics custom two-tone stock finish.

Re-barrelled Anschütz with a .17AK Improved cartridge, Pac-Nor 1 in 9in twist barrel, Hydrographics camouflage-wrapped stock and MAE sound moderator.

DOLPHIN GUN COMPANY

www.dolphinguncompany.co.uk
01507 343898
The Dolphin Gun Company was formed by Mik Maksimovic and Pete Hobson, who both come from engineering backgrounds and are well known as competitive F Class shooters,. The Dolphin Gun Company was initially set up to help their competitive shooting careers. Their purpose is to provide precision extreme rifle accuracy at a good price and in a realistic time frame

Dolphin also makes an aluminium-framed chassis that can be inlet to a variety of actions. These stocks are all made in-house on one of the many CNC machines. They are fully adjustable, including the pistol grip, which can be adjusted for length of pull, with a choice of three types of forend: short tactical style, long F/TR style and a Benchrest F Open forend. They are also available in a choice of anodized colours or coated in-house with Duracoat in the client's choice of colour. Dolphin maintains a large stock of barrels, actions, stocks, triggers and scope mounts, as well as all the ancillary paraphernalia that goes into making a custom rifle. One of Mik's in-house specialities is barrel fluting, with a selection of straight, spiral, cross-over and interrupted fluting.

Dolphin's own action, the CST DGC, is entirely designed and machined in-house and

Dolphin Gun Company's Nesika F/TR rifle features a Nesika K action, Jewell BR trigger, fluted bolt and Bartlein 30in, 1 in 7in twist barrel, .223 AK chambered for the Sierra 90 grain bullets. This is all bedded to a Dolphin modular stock and supported by a Dolphin Trakker 1 bipod.

This Nesika F/TR rifle, featuring a Nesika K action, custom bolt knob and fluted bolt, is available with a choice of barrel makes and chambering. Either a Jewell BR or HVR trigger may be fitted to a Dolphin modular stock in the colour of your choice.

Dolphin also caters for extreme-range rifle shooters with this Deviant action 7mm Rem Magnum rifle, fully pillar bedded into a Joe West laminate stock.

Mik from Dolphin produces some of the best fluted barrels in Britain with superbly executed styles and finishes.

is as good as it gets. It has been engineered from the ground up and has many innovative features to provide the most accurate shooting platform obtainable in a custom rifle. As well as re-barrelling, threading and all the usual gunsmithing services, Dolphin offers its own styled custom rifles in the form of the Tactical, FT or sporter.

STEVE KERSHAW FIREARMS

www.stevekershawfirearms.co.uk
01430 430553
Steve Kershaw, based in Howden, East Yorkshire, is a registered firearms dealer with more than thirty years of firearms and hunting knowledge, ranging from pest control through to deer

An ultra-short 6.5 × 47L chambered Walther barrel with 1 in 8in rifling twist and 18in length, fitted with a MAE Scout sound moderator. It is fitted to an AI AX folding stock and the action is a Stiller Precision Predator Tac 30 with custom blacked surface and tactical bolt handle.

Fully blueprinted Remington 700 action chambered for .243 AK Improved with a Krieger cut rifled 1 in 8in twist barrel fitted with an MAE Scout moderator, bedded to a McMillan stock with full pillar bedding.

Another Predator actioned rifle, this time bedded to an AI AICS stock and chambered for the 6mm BR round. The barrel is a 1 in 8in twist Lothar Walther and the action is fitted with a set of Talley scope mounts.

stalking and hunting abroad. He can build, feed and supply all the kit necessary to keep custom rifles performing correctly. All rifles are built with extreme accuracy in mind, including precision rimfires, hunting, varminting, tactical and F Class target rifles. Each is made to the customer's requirements in any configuration and the same detailed attention is paid to a re-barrel as to a ground-up, custom-built rifle. A good choice of barrel, action and stock types is always available. Steve regularly uses either Lothar Walther or Krieger blanks, which can be ordered in any profile, fluting length and barrel twists, and can be made of either chrome moly or stainless steel. In-house chambering is done with any standard reamer cartridge, but wildcats can also be specified if you feel adventurous. All are shot and sighted in and load data provided before collection by the customer to ensure 100 per cent satisfaction is maintained.

ANGLO CUSTOM RIFLE COMPANY

www.anglocustomrifle.co.uk
07718 911158
Anglo Custom Rifle Company was established by Gavin Haywood in 2008 to manufacture specialist weapons for discerning shooters. Whether it is re-barrelling your old rifle or building a top-line competition rifle, Gavin can sort a rifle to perform at its best. Anglo can make a bespoke rifle but also offers a range of off-the-shelf rifles, including the lightweight and titanium models, fitted with carbon fibre stocks. These have the very best modern 5R barrels sourced from Border Barrels and Impala actions made from either stainless steel or titanium. The second custom option is the sporting rifle, with a stock selected from ranges by McMillan, Robertson, Manners, Accuracy International and HS Precision, and a finely tuned action by Impala Precision. If preferred, a blueprinted customer's action, or one by Stiller, Stolle, Borden, Lawton or BAT may be fitted. Barrels from makers such as Krieger, Border, Bartlein and Hart are then precision profiled and chambered for any target, varminting or hunting need. Gavin can also arrange for a classic using custom parts bedded into a walnut stock, oil finished and superbly chequered for a custom English-style wood-stocked rifle.

An Impala SR7 action in .260 Ackley Improved with a 1 in 8in twist Sassen barrel from Birmingham, Jewell trigger and custom stainless floor plate. The Cerakote metal finishing is by Richard Harvey in Northampton.

The wood stock is of handmade Exhibition grade burr walnut with hand-rubbed oil finish and superb chequering done by Cherrie Abrahams. There is a full aluminium and Devcon bedding for the action.

An Impala SR7 action with a Shilen stainless steel barrel chambered in .260 Remington, with an invisible thread protector at the muzzle with an 11 degree target crown. The stock has a custom laid-up full naked carbon fibre monocoque and has a right-hand 15.5in pull length. This makes it an extremely lightweight and accurate sporter.

This precision rifle with an Impala SR7 action features an Oberndorf floor plate, Jewell HVRT trigger with aluminium pillars and full thickness Devcon bedding.

Another in-house Impala SR7 action in .308 with a Rock-Creek 5R barrel carbon wrapped by K.K. Jense at Jense Precision in the USA. The stock is a Carbon PSE Tac unit with adjustable cheekpiece. The whole rifle weighs less than 3.2kg (7lb) and includes an aluminium bottom metal accepting AI magazines. It is a 1200-yard capable rifle that can be carried all day on the hill.

The Impala SR7 action is superbly made and has a Rem 700 footprint that can take all manner of aftermarket items to put together your dream custom rifle.

RIFLECRAFT LTD

www.riflecraft.co.uk
01379 853745
Andrew Evan Hendricks comes from a military background and as such knows how a rifle works. After careers in the army and as an agricultural chemist, he decided to follow his dream. Riflecraft builds affordable custom rifles for clients who use rifles for a living and for recreation. Andrew offers precise custom rifles built to a customer's exact customer requirements and also offers a full training programme for long-range shooting, reloading and advanced rifle handling, all provided in-house on purpose-built ranges. Riflecraft has a fully stocked

A .308 Win Tactical using a 20in, 1 in 10in twist barrel that is Marine-Tex pillar bedded into a McMillan A5 adjustable stock with a Hardy sound moderator and Cerakote finish.

showroom in Harleston, Norfolk, and can supply custom rifle parts, rebuilds, full custom or re-barrelled rifles. The firm also has in-house gunsmiths and a professional Cerakote coating service. Andrew's typical custom rifles cover the gamut from .22 rimfire benchrest models all the way to one-off special sporter-orientated pieces and military specification sniper rifles.

A Remington SPS action and TET scope rail, fitted with a Timney trigger and KRG tactical bolt knob and Badger Ordnance DM floor plate to accept MDT magazines.

This .22LR BR50 Stiller Lone Star single-shot action is fitted with a Border 1 in 16in twist, S/S S/M fluted, 24in barrel and Marine-Tex pillar bedded to a McMillan rimfire Magnum stock for highly accurate 200 yard rimfire shooting.

The precision Stiller Lone Star action is fitted with Jewell trigger, custom-made trigger guard and Stiller scope rail.

Riflecraft 6.5 × 284 fox or deer rifle on a blueprinted Rem 700 LA RH action with Precision recoil lug (.250in thick), bedded with Marine-Tex into a H&S precision stock with a Hardy sound moderator and custom Cerakote finish.

Blueprinted Remington action with a Pac-Nor 1 in 8.5in twist, S/S S/M 2in barrel with a Badger Ordnance rail and Tactical bolt handle, and Timney trigger.

6.5 × 55mm chambered Krieger S/S S/M fluted, 1 in 8in twist barrel on a Sako SS action (trued) and bolt fluted with an all-metalwork Titanium Cerakote finish (FDE/White/Black 'Splatter') in a McMillan Sako varmint stock.

VALKYRIE RIFLES

www.valkyrierifles.net
07889 388378
David Wylde started working in a gun and tackle shop repairing all sorts of rifles, shotguns and air weapons. He has more than twenty-five years' experience in gunsmithing. He is a multiple gold medal winner at national and international level in Gallery Rifle, F class, Civilian Service Rifle (CSR) and practical rifle, and holds four current British records. He also knows how to build custom rifles as well as shoot them.

David can make anything from a deer stalking classic rifle to a long-range varminter, a straight-pull military specification or a one-off extreme custom rifle. He is an avid CSR shooter and produces rifles for this type of discipline that are capable of one-hole groups at 100 yards. He also has a real skill for producing special paint finishes and he excels at Cerakote finished rifles of singular or multi-coloured texture. He has recently been appointed the sole Accuracy International factory-trained and approved service centre in England and as such is authorized to carry out work on these prestigious rifles.

Materials and components necessary to build and complete a custom rifle of your choice are usually in stock or can be ordered to meet exactly the correct barrel action and stock needed for a custom project. All work such as barrel chambering, threading, blueprinting, stock bedding, Cerakote finishing and machining individual items, such as scope mounts and muzzle brakes, are carried out in-house, as well as specialist paint finishes.

.308 Win chambered Thor action with a 1 in 12in twist, Sassen Engineering 20in barrel with helical fluting, Timney trigger and teardrop bolt handle, bedded with Devcon into a Chaparral carbon composite stock and finished in Titanium Armour Black Cerakote.

David Wylde's signature rifle, his own .22BR Valhalla actioned, single-shot, long-range tack driver. This rifle incorporates everything in a true custom rifle with precision-made parts and a superb finish.

Fitted with a Bartlein 1 in 8in twist barrel and chambered in .22BR to shoot heavier bullet types, this Valhalla actioned rifle is fully bedded with aluminium pillars and Devcon bedding material into a Robertson varminter stock. The finish consists of a black base overlaid with silver marble and candy apple red, which has a life of its own.

The Valhalla action is highly polished with cam locks on the bolt, which has superb barley twist fluting. A faceted bolt shroud, Jewell BR trigger and Sunny Hill stainless steel floor plate complete the finish.

6mm BR rifle with Surgeon action and 1 in 8in twist, Border cut rifled barrel with straight fluting, 28in in length and with a pepperbox muzzle brake. The stock is an XLR modular system with Ergl pistol grip and AI magazine all, finished in David Wylde's twenty-coloured custom Cerakote finish.

David also specializes in CSR (Civilian Service Rifle) type rifles. This has a DPMS upper and Double Star lower receiver and a Krieger 1 in 9in twist, .223 Wylde 14in barrel. The stock is a Magpul UBR with Magpul magazine, Christensen forend and RPA muzzle brake, all coated in Patriot Brown Cerakote finish.

David Wylde also blueprints Remington actions when not using his own Thor or Valhalla actions. This is a Rem blueprint 260 Rem-chambered, Sassen Engineering stainless steel barrel of 1 in 8.5in rifling twist and 26in total length. It is bedded into an AI CSAP stock and has a custom bolt handle. It is finished in Graphite Black Cerakote.

MCKILLOP ENGINEERING

0118 933 3100
Neil Mckillop learnt his gunsmithing skills as an armourer in the armed services working on all manner of weaponry. He later turned his technological and engineering experience to work in the racing car sector, primarily with composite fibre technology. As with many of the custom makers here, his hobby soon became his passion and then his job. Neil has a well-stocked machinery shop on the outskirts of Reading. His day-to-day workload consists of re-barrelling rifles for local deer stalkers or pest control companies, or threading barrels for fitment of sound moderators. He also builds custom one-off rifles for clients who require extreme accuracy for shooting disciplines at the nearby Bisley shooting ranges, for long-range varminting or for clients wanting a

Full-blown F Class extreme-range custom rifles are Neil's usual fare. Here is a .284 Win 0.315in neck diameter chambered BAT M action with 32in No. 4 profile Bartlein barrel with 1 in 9in rifling twist. The stock is a Joe West grey laminate with twin adjuster, cheekpiece, pillar and Devcon bedded, and an F class thumbhole design.

rifle capable of superior accuracy. Neil builds many custom rifles for F Class shooters who require a competitive rifle at the highest level and his repeat business shows he knows how to build a winning rifle. Being a machinist, he is also able to make custom items from scratch, such as muzzle brakes and bolt handles, as well re-barrelling customers' rifles or bedding a new stock with aluminium pillars and Devcon bedding material.

The BAT M action is perfect for a precision long-range rifle and the Jewell HVR trigger makes the most of the accuracy potential.

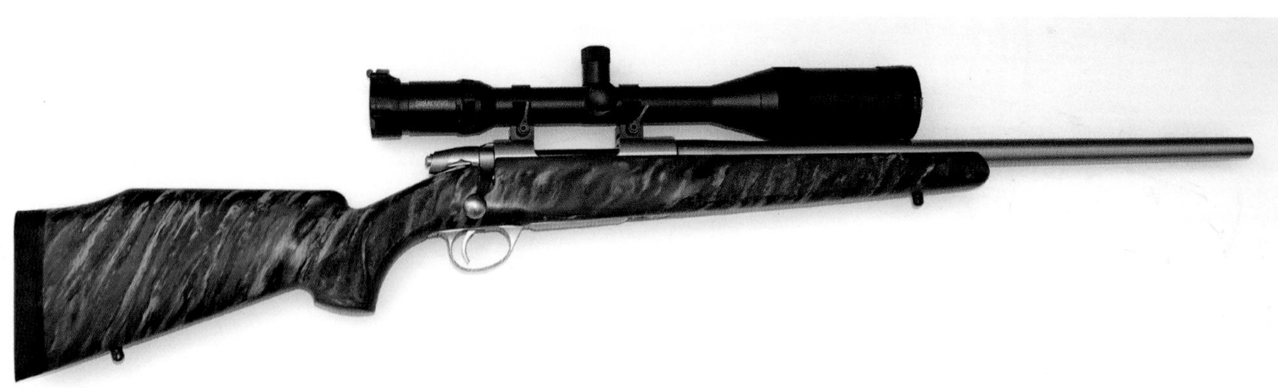

Neil also re-barrels customers' rifles. Here is a Sako 75 action rebarrelled with a Pac-Nor 1 in 8in twist, four-groove 22.25in barrel in .243Win with a 0.276in neck. It has an invisible muzzle threaded for a sound moderator and a vapour-blasted finish. It is pillar and Devcon bedded into a McMillan Sako varmint-style stock with black/green/beige marble finish.

DEVON CUSTOM RIFLES

www.devoncustomrifles.co.uk
01548 856742
Based in Churchstow, Devon, Devon Custom Rifles (DCR) is a newcomer to the custom rifle scene, but its CNC machine-based workshop has the capability to produce actions, triggers, custom parts and now rifles. Gary Alden, the managing director, and John Davies, the firearms manager, aim is to provide British-made rifles using parts they produce themselves, such as actions, triggers, bolt handles, stocks, scope rails and barrels sourced from Sassen Engineering (Border Barrels).

DCR's actions are now used by many custom rifle builders in Britain and rebadged under their own names. Actions can be built for all cartridge sizes and bolt and port orientations. The triggers are also built in Britain and are readily available, overcoming the problems of sourcing kit from abroad. DCR will re-barrel a customer's rifle, if necessary, and retro-fit their own trigger units, but they are continuing to broaden their range of custom rifles to provide value and accuracy in one home-grown package.

LEFT: *This is a full blown custom special: DCR multi-shot, right port and right bolt action with DCR 20 MOA scope rail. The barrel is a Pac-Nor 27in Super Match Grade fluted Remington varmint profile with screw-cut muzzle. The stock is a PSE E-Tac and fully pillar bedded.*

LEFT: *The DCR action has a DCR pineapple bolt handle and DCR hybrid trigger of 8oz with Absolute Accuracy Mag Con bottom metal and five-round AICS magazine.*

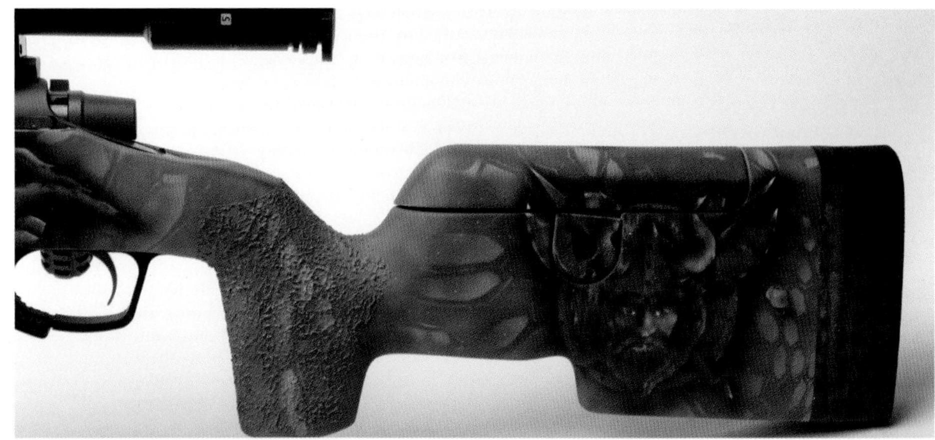

The superb stock finish is custom airbrushed by Gary Lowe. The rifle is finished off with a Calibre Innovations sound moderator and Minox ZP5 scope.

Useful Addresses

Stocks	*UK suppliers*
McMillan and Robertson	Jackson Rifles www.jacksonrifles.com; 01644 470223
Bell and Carlson www.bellandcarlson.com; 001 620 225 6688	
H–S Precision www.hsprecision.com; 001 605 341 3006	
Christensen Arms www.christensen arms.com	Dolphin Gun Company www.dolphinguncompany.co.uk; 01507 343898
Manners Composite Stocks www.mannersstocks.com	MarchScopes www.marchscopes.co.uk; 01293 606901
Accuracy International Ltd www.accuracyinternational.com	Sporting Services www.sportingservices.co.uk; 01342 716427
CCL oil finishing kit	Uttings Ltd www.uttings.co.uk; 01603 619811
Chapparal Composites Ltd	www.chaparralcomposites.com; 07540 808980

Actions	
RPA Defence Ltd	www.rpadefence.com; 0845 880 3222
Stiller's Precision Firearms www.viperactions.com	TWG Technical Services 07766 720404
BAT Machine Co., Inc www.batmachine.com; 001 208 687 0341	
Borden Accuracy www.bordenrifles.com	Precision Rifle Services www.precisionrifles.com; 01807 580422
Nesika www.nesikafirearms.com	Dolphin Gun Company www.dolphinguncompany.co.uk; 01507 343898
Gerry Geske 001 406 822 4917	
Barnard Precision www.barnard.co.nz	Suppliers include www.normanclarkgunsmith.com and www.dolphinguncompany.co.uk
CST DGC Dolphin	Dolphin Gun Company www.dolphinguncompany.co.uk; 01507 343898
DCR	Devon Custom Rifles www.devoncustomrifles.co.uk; 01548 856742

Mayfair	Mayfair Engineering www.mayfairengineering.com; 01379 871510
Valkyrie	Valkyrie Rifles www.valkyrierifles.net; 07889 388378
Impala Precision	Anglo Custom Rifle Company www.anglocustomrifle.co.uk; 07718 911158

Barrels

Sassen Engineering (Border Barrels)	Border Barrels Limited www.border–barrels.com; 0121 359 7411
Pac–Nor, Krieger, Bartlein, Shilen, Lilja	Available on request from all good custom rifle makers
Lothar Walther www.lothar–walther.com	
MAE moderators	JMS Arms www.jmsarms.com; 01444 400126
Scorpion reamers	Scorpion Tooling UK Limited www.scorpiontooling.co.uk; 01453 549321
Pacific Tool & Gauge reamers www.pacifictoolandgauge.com; 001 541 826 5808	

Triggers and Safeties

G. Recknagel e.K. www.recknagel.de	Alan Rhone Ltd www.alanrhone.com; 01978 660001
DCR	Devon Custom Rifles www.devoncustomrifles.co.uk; 01548 856742
Jewell	Jackson Rifles www.jacksonrifles.com; 01644 470223
Timney Triggers www.timneytriggers.com CD Universal	Riflecraft Ltd www.riflecraft.co.uk; 01379 853745 Jackson Rifles www.jacksonrifles.com; 01644 470223
Cadex and Trigger Tech	Sporting Services www.sportingservices.co.uk; 01342 716427
Rowan Engineering	www.rowanengineering.com; 01295 251188

Scope Mounts and Sights

Recknagel and Ziegler ZP	Alan Rhone Ltd www.alanrhone.com; 01978 660001

Burris
Leupold

GMK Limited
www.gmk.co.uk; 01379 871510

Sako/Tikka Optilock
www.tikka.fi

Viking Arms Ltd
www.vikingarms.com; 01423 780810

Tier One

Evo Leisure Limited
www.tier–one.eu; 01924 404312

Talley Manufacturing Inc
www.talleymanufacturing.com; 001 803 854 5700

Conetrol
www.conetrol.com; 001 830 379 3030

EAW
www.eaw.de

Sentry Trading
www.st.uk.com; 01420 300123

Warne Manufacturing Company
www.warnescopemounts.com

Sportsman Gun Centre Ltd
www.sportsmanguncentre.com;
01392 354854

JP Smithson
www.smithson–gunmaker.com; 001 801 224 2041

Yukon Advanced Optics (Pulsar)
www.pulsar–nv.com

Thomas Jacks Limited
www.thomasjacks.co.uk; 01789 264100

Trimmings

Paul Richins

Spirit Carvings; 07843 118208

Recknagel (custom parts)

Alan Rhone Ltd
www.alanrhone.com; 01978 660001

Prestige Custom Coatings Ltd

www.prestigecustomcoatings.co.uk;
0800 476450

Cerakote Ceramic Firearm Coatings
www.cerakoteguncoatings.com

Suppliers include

www.valkyrierifles.net and www.riflecraft.co.uk

Hydro Graphics

www.hydro–graphics.co.uk;
01904 778188

Feeding a Custom Rifle

Midway

www.midwayuk.com

Norman Clark Gunsmiths

www.normanclarkgunsmith.com;
01788 579651

Reloading Solutions Limited

www.reloadingsolutions.com;
01865 378200

QuickLOAD
www.quickload.co.uk

JMS Arms
www.jmsarms.com; 01444 400126

Index